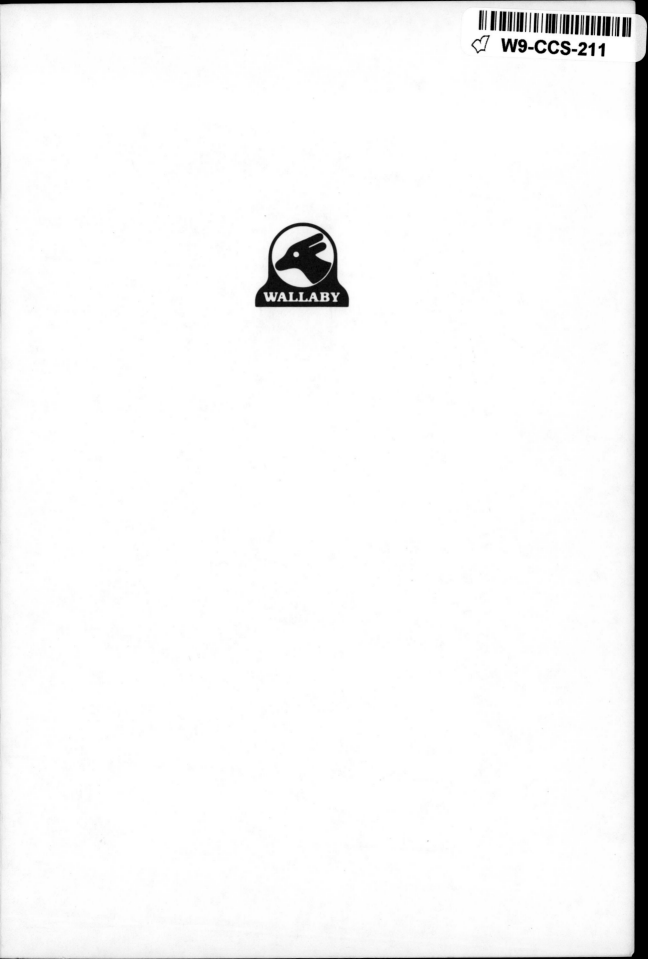

WALLABY

THE JOY OF TOUCH

RUSS A. RUEGER, PH.D.

Photographs by Trudy Schlachter

A WALLABY BOOK
Published by Simon & Schuster
NEW YORK

DEDICATION

to little lizzie,
whose cuddly warmth helped me through many long hours of writing and typing.

Published by Wallaby Books
A Simon & Schuster Division of Gulf & Western Corporation
Simon & Schuster Building
1230 Avenue of the Americas
New York, New York 10020
WALLABY and Colophon are registered trademarks of Simon & Schuster
First Wallaby Books Printing September, 1981
Manufactured in the United States of America

10 9 8 7 6 5

ISBN 0-671-42469-6

This book is literally the product of a dream. I first came upon the idea of writing about touch in 1973, when I had an article on cuddling published in a major men's magazine. I followed this up with a 1976 essay on tickling for another men's mag. In 1979, while scraping by as an underpaid law clerk, I started seriously thinking about expanding the cuddling and tickling ideas into a book. I knew the concept connecting them was touch, but I couldn't figure out what else should go in a touch book.

I thought about the problem for months, and got nowhere. In late October, 1979, I went to sleep still thinking about it. But then I had a vivid dream. I was sitting at my typewriter, and some unknown man was dictating to me the chapters that belong in the touch book. He told me to include massage, Shiatsu, healing, and other subjects. He also said the book was a "viable commercial possibility." I then woke up and went right from my dream typewriter to the real one, and typed out the chapter titles that had been given to me.

A few days later, I casually mentioned my idea to an editor friend, and he took me in to see his publisher. I was stunned when the publisher spontaneously described the idea as a "definite commercial possibility"—almost the exact dream phrase!

Since then, it's been enchanting to watch the dream come true. Through each stage of the project—writing, typing, editing, proofing, photography—the dream has increasingly materialized into concrete reality until now, the finished product. I don't know whether the unknown man in my dream was from the depths of my subconscious or some more divine source. I do know that he played a crucial role in the creation of this book. And I want to acknowledge that contribution.

I also want to acknowledge the contributions of the many flesh and blood people who helped with this project. Particular thanks for editorial assistance go to Nils Shapiro and Eric Protter of Gallery Press, and Gene Brissie of Simon & Schuster. My talented and perceptive photographer, Trudi Schlachter, did a truly inspired job putting a visual frame on the words in this book. Numerous other people contributed ideas or explained techniques, and they have been acknowledged in the text itself.

There's a final group of people who contributed in an intangible, but very real, sense. These are the many wonderful women in my life, starting with my mother, who taught me the value of physical closeness and tender loving care. Some of them specifically urged me to write about touch. They all added to my appreciation of its soothing, healing qualities. I thank all of them for expanding my awareness and sensitivities. They know who they are.

CONTENTS

PART I: THE TOUCHING DRIVE

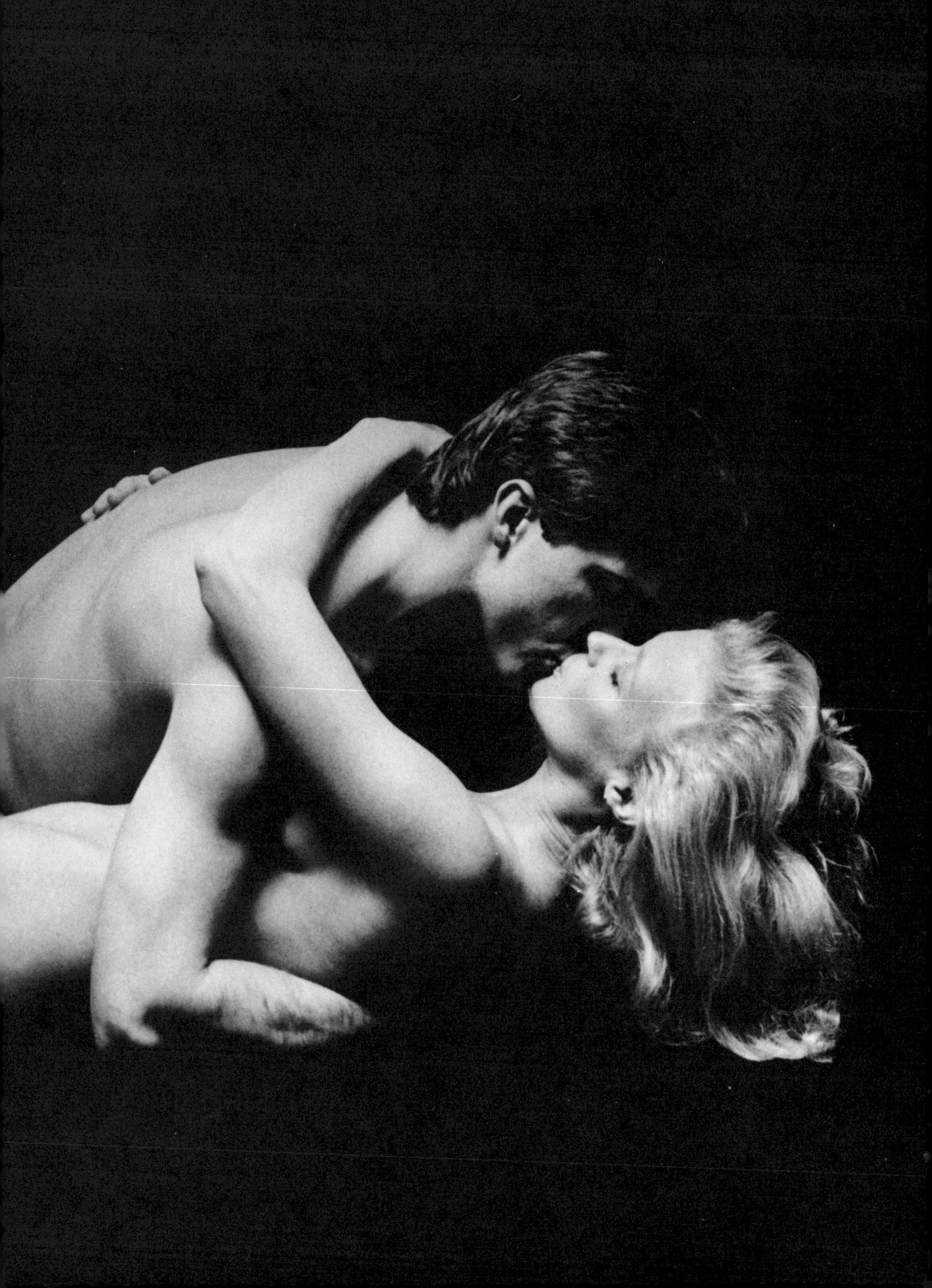

CHAPTER ONE: THE SHUNNED SENSE

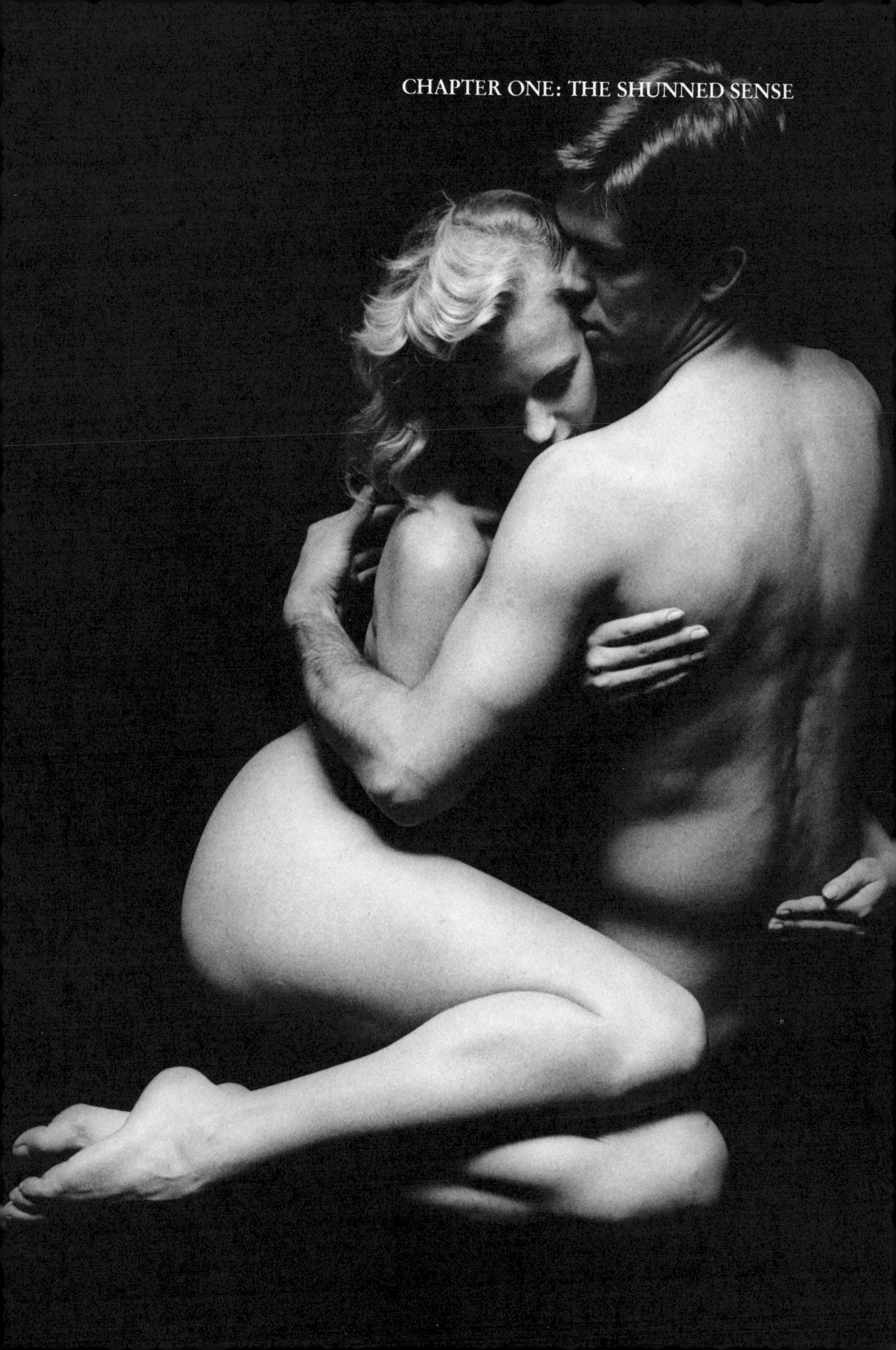

Billions of dollars are spent each year satisfying four of the five human senses. Entire industries have sprung up around one or more of them: TV and film serve sight; radio and records pitch to sound; supermarkets and restaurants satisfy taste, and cosmetic companies cater to smell. In contrast, touch has been very much left out in the cold.

Some services on the market do enhance touch, but even they downplay the touch aspect. Skin care products, for instance, put as much stress on the benefits of "healthy-looking" skin as they do "healthy-feeling" skin. Ads for shower-massage gadgets show people lavishing in delicious, tingly skin sensations. But they constantly remind you about the practical connection with bathing: "Why just turn on the water, when the water can turn you on?"

A *true* touch industry would show touch enjoyment *apart* from anything else. In the rare cases where this does occur, the product or service loses respectability. For example, both vibrators and massage have become linked in people's minds with sex play and prostitution. A purely touch-oriented vibrator would be viewed by most as a thinly veiled sex product. Similarly, a legitimate, licensed masseuse finds herself lumped together with ladies who give brothel-style "back rubs."

Obviously, society accords contact activities a very special treatment. Touch has become so psychologically connected with sex that people transfer the prohibitions of sex to touch. Naturally, sex does entail an intense touch experience, but the two are hardly interchangeable. Nor is sex the sole type of intimate contact. Conversely, there's more to sex than touch. Sex also involves the senses of sight, sound, smell and taste. Would anyone seriously suggest putting blinders on people in public because the eyes always have the potential to leer at nude skin? Or perhaps banning public eating because it could suggest oral sex play? Silly as these suggestions sound, touch is not treated all that differently. No reputable industry caters solely to the shunned sense because touch has been shackled to the bedroom door. The less we see it outside that spot, the more comfortable we seem to feel.

We also find that public contact with others is strictly limited. Whatever *is* allowed has been filtered into ritualized gestures like handshakes and pats on the back. Touch restrictions generally grow greater as civilization advances. For example, in a crowded American city, the slightest accidental brushing of two strangers brings forth profuse apologies. Is this fleeting contact anything to really feel "sorry" about? The lack of aggressive intent is usually rather obvious. What is truly threatening about the momentary brush is that it violates the "no touch" rule among strangers. The elaborate apologies publicly demonstrate that the touchers genuinely pay homage to the rule. Only circumstances clearly beyond their control caused them to break it.

Unfortunately, this straitjacket approach to touch binds our growth and development. If you blindfold someone for most of the day, you could hardly expect his sight to work right the rest of the time. The contrast would be too strong, and his vision would likely be impaired even when the blindfold was off. The same is true for touch. You can't expect people to function all the time in public under strict "don't touch" rules, and then let it all hang out in private. The constant prohibition literally gets under your skin, causing contact inhibitions even with close friends. Touch neuroticism can become so deep-seated that you can become "out of touch" with basic bodily needs. As we shall see later in this book, some people grow ill at the very thought of being touched; theirs is a real problem, whether or not they're aware of it. The special touch techniques we shall describe can help reverse this process.

The contact prohibitions are particularly troublesome because they interfere with one of our most fundamental biological functions. Touch is the first sense to develop, since the other four are specialized outgrowths of tactile tissue. Touch is the primal human experience. Our lives begin immersed in the enveloping nurturance of the womb. The womb-state gives us the most encompassing embrace we'll ever know — total enclosure inside the body of another. The fetus floats blissfully in this dark, fleshy nirvana, and there's little doubt that this full-touch experience becomes embedded somewhere in the deeper recesses of the psyche. Man builds enclosures wherever he lives, and shelter is a key part of man's drive for security. Psychologists often interpret dreams of entering a room as symbols for penetrating a womb. The room may well represent the womb. However, the dreamer may not be seeking sex, but the exquisite womb-enfoldment impressed in his subconscious memory.

While touch gets its take-off in the womb, it also plays a central role in early growth. The massive contractions of the uterus in labor stimulate the newborn's digestive and respiratory tracts. Essentially, this intense contact throws the switch to start the baby's life-support systems. The traditional slap on the baby's backside serves a similar purpose. Once born, the baby's early sensations will be overwhelmingly tactile. The "proximity" senses of touch, taste and smell are primary in the early years, while the "distance" senses of sight and sound mature later.

The infant's initial learning experiences come through contact. Through the close care of the mother, the world unfolds its mysteries through touch. Love, security, nurturance, trust, affection, motion and time all introduce themselves through the maternal embrace. The baby's basic reactions to other people are formed from his early skin contacts. The amount of tender, loving care (TLC) he receives in his first few years directly influences how much TLC he can return later in life. Touch between people is actually a form of non-verbal communication which can expose the toucher's primary attitudes. The hand cannot lie as easily as the mouth. An infant handled in a cold, distant way will find it tough to resist the message that the universe is a cold, distant place.

Contact is also crucial for the infant's experience with objects. A baby is rarely satisfied with just gazing at something interesting. Instead, he'll draw it close to him, often putting it in his mouth, one of the most tactile spots of all. He uses touch to define the boundaries between his body and everything else. Touch lets him learn that something impinging upon his skin is not a part of himself. This, of course, is the beginning of self-awareness.

As the child grows up, his contact quotient is drastically cut back. Like pets, it's okay for almost anyone to touch a baby or toddler, but it becomes a problem after that. Well before puberty, parents begin to read sexual overtones in caresses which were once commonplace. At some indefinable age, the no-touch rules take over. Unfortunately, the desire for touch is never outgrown. There's nothing "childish" or "infantile" about the craving for warm, nurturing contact, although society often treats it so.

Actually, warm body contact occupies a special place in our hearts. It's very much like returning home — our very first home, the womb. No matter how many years an adult has been on his own, the visit to parents still has special significance. The home-cooked meals and other memories of childhood create a warm glow inside. Much the same is true of the hearty embrace. It helps us relive our early experiences of being loved and cared for. And that's something most of us would not *want* to outgrow.

Our ape cousins, as well as less-developed human cultures, are not as "out of touch" as we are. Chimpanzees, monkeys and other apes get great delight from social grooming of the hair. This may help rid them of insect pests, but the touch gratification is obvious. In fact, the invitation to groom is a friendly gesture and a means for forming strong social bonds between apes. The invitation to groom looks a lot like a bow, which is still a form of greeting for people. Some less-developed *human* societies also engage in hair-grooming, and are generally less uptight about touch. Such societies, of course, are smaller and close-knit, so just about everyone knows everyone else. In a real sense, they're all one big family. Here, familiarity breeds contact. Even in Western society, touch restrictions are considerably relaxed among family members. Unfortunately, the West is moving further from family patterns and more toward anonymity for all. Instead of treating more outsiders like family, we treat more family members like outsiders.

Family life in America was once a huddle of aunts, uncles, cousins, grandparents, nephews, nieces, etc. —the "extended" family. In those days, lots of hugs came from all directions. If you were "family," you were touchable. Nowadays, the family unit has dwindled down to the unstable "nuclear" core of parents and offspring. Contact with other kin has been drastically reduced, and the former hearty hugs are now quite subdued. When you see a relative once a year or so, there's not much motivation to treat him any differently than you would a stranger. Our lives become increasingly populated by strangers and others who are functional strangers —like business "contacts" (hardly an apt description) and out-of-touch kinfolk. TV exacerbates this tendency to treat people in a cold, detached way. We're confronted daily with dozens of TV-denizens who share everything with us but their flesh-and-blood presence. We vividly see their images, hear their words, share their sentiments, but they don't "touch" us deep inside. Only warm, living contact can do that.

The constant exposure to tube people actually has a narcotic effect on our normal sensibilities. We see them being shot up, hear their screams, but don't truly feel the violence, so we become more or less immune to it all. Aggression and other wrenching events watched on TV just aren't close enough to real life, so we tend to discount them. They only hurt the "other guy."

Unfortunately, we sometimes carry this effect into real life. After all, if we've seen thousands of TV muggings, why should we be affected by another one in the street? TV helps us to overlook the true human significance of such a tragedy. How different it might be if TV could make us viscerally experience what we now only see and hear. "Feelevision" would let us intimately penetrate the tragedies and triumphs which are now just noisy, abstract images. Being able to feel the action would drastically dwindle our present detachment. Touch is undoubtedly our most humanizing sense.

While the no-touch rules weave an ever-expanding circle, our language continues to reflect the true significance of touch. The words "touch" and "feel" and their variations are among the most lengthy entries in the dictionary. There's a seemingly endless list of metaphors relating to touch. An insensitive individual is said to be "thick-skinned," "heavy-handed," or "tactless." Such a person gets "under the skin" because he "rubs us" the wrong way. More sensitive sorts are "tactful" or "soft touches." More recent variations include being "stroked" for good treatment and "picking up good vibrations" for encouraging messages from the opposite sex.

It's also intriguing the way the words "touch" and "feel" are connected. In one sense, they mean the same thing: to touch something *is* to feel it. In another respect, to "feel" a certain way ("I feel happy") means to experience an emotional state. It's as though the body were being enveloped by or wrapped up in the emotion. Similarly, to be "touched" by something means to be deeply moved by it. Apparently, our language almost equates touching and feeling with emotion. Obviously, touch is a potent tool. Fearing its power, society has fenced it inside an intricate set of prohibitions. Touch itself has become almost "untouchable."

Have you ever watched the way two animal strangers approach each other? Like human strangers, they watch each other tentatively at first, cautious for any threat signals. If neither threatens, they stare for a while, then usually go right up and touch. They may sniff, brush, rub, or inspect each other's hind parts, whether or not there's sexual interest. Friendly curiosity naturally leads to the desire to make close contact.

Contrast this with human strangers. In most cases, a brief eye acknowledgment is followed by rigidly staring ahead (or looking down) as the two pass. If there's special interest or attraction — such as sexual interest between opposite-sex strangers — the eye contact may be prolonged, and may even be accompanied by a smile or greeting. However, even if two passersby were absolutely *made* for each other, if they were drawn like magnets and every instinct cried out, "I want to *know* this person," they're hardly likely to start fondling. In fact, even their speech will downplay their true interest. If they can say anything at all — and most people get tongue-tied in such situations — it might be a subdued pitch for a date.

The physical attraction felt by the pair would mostly remain concealed in their conversation. True communication would come non-verbally through expression, gestures and postures. The pair may lean towards each other, thrust their hips forward, or maintain close eye contact. These movements symbolically make the point that the animal strangers made directly through touch — "I like you," "You're sexy," etc. Problem is, symbolic gestures are ambiguous. They can be interpreted in a number of ways. In 9 out of 10 cases, magnetically drawn strangers, no matter how highly motivated, are *not* going to get to know each other through an anonymous street encounter. The social restraints really stack the deck against it.

Fact is, even sex and romantic attraction, among the most potent human drives, can't make much of a dent in the touch taboo against strangers. In a real sense, strangers form a class of untouchables. The touch taboo is the body armor we wear to ward off potential intruders. Strangers could be dangerous or crazy. Who knows what's lurking in their silent skulls? Even seemingly nice strangers can be upsetting, drawing us away from our present loves, waylaying us from carefully laid plans, opening up new, unimagined vistas. Strangers must be kept away so we can keep our lives safely in their narrow, controlled grooves. The touch taboo takes care of the problem by prohibiting most spontaneous contact with strangers. Actively bucking this rule would be asking for a quick route to a looney bin or jail cell.

The "don't touch strangers" dictum may be the ancestor of all touch taboos. It covers two concerns closely connected with contact: sex and violence. By banning spontaneous sexual encounters, it separates man from beast. The taboo allows mating to become safely channeled into proper social rituals like courtship and marriage — which makes it a lot easier to pin down the father of a female's child, and maintain family responsibility. It also helps keep marriages in the right class and power relationships. We'd have a rather different society if a poor farmboy were permitted to express spontaneous affection for a rich heiress.

The taboo also allows the state a monopoly on legitimate force. For example, police officers, strangers though they may be, are exempt from the touch rules in their official roles. They can grab us, lead us, nudge us, and direct us without our

feeling violated. Only when they become grossly excessive ("brutal") do we raise a fuss. By prohibiting us from pushing each other around, the taboo helps society maintain control. The burden of defending ourselves becomes the state's responsibility.

Unfortunately, sex and aggression are only the extreme ends of the touch scale. Sex is close to the positive pole, while violence lies on the negative pole. The touch taboo does an admirable job at both ends, but it also wipes out everything in between. We respond to an accidental brush almost as though it were active aggression. Our laws also don't distinguish very well between degrees of touch. For example, the civil law concept of battery generally covers any "unwanted touching." This encompasses both a gentle kiss and a punch in the nose.

Close cousins of the stranger touch taboo are the various sex-connected contact restrictions. Many of these trace back to Christianity's traditional opposition to bodily pleasures. The body was seen as a base, inferior part of man's existence, "mired in the mud." The true mark of a man was his spirit or soul. Denial of bodily drives was somehow supposed to temper and refine the spirit. From the beginning, "To dust thou art and to dust thou shalt return" has reminded Christians that their bodies were inseparable from the transient material world. The dwelling place of the spirit, however, was in heaven, so it was sinful to pay homage to things of the world.

Therefore, bodily pleasures served the world, and were to be kept to a minimum. Sex was excusable only within marriage *and* for procreation. Since sex is impossible without touch, a sex/touch taboo emerged. You could never tell what that first touch would lead to, so it was best to ban it whenever possible. American Puritan traditions have been heavily influenced by these ideas, perhaps more so than other Christian countries. Only Germany and Britain (like the U.S., Protestant nations) are more reserved about touching than America. Countries with non-Protestant Christian traditions like Greece, Russia, Latin America and Mediterranean Europe are much less uptight.

The sex/touch taboo comes in many shapes and sizes. The number one pain-producer has to be the same-sex taboo. The big fear here is homosexuality.

After all, gays do touch each other, so any sustained same-sex contact must be a sign of gayness. Like many touch rules, the same-sex taboo differs for males and females. Women are permitted to kiss and hold hands, while this is totally off-base for men. The most involved male – male contact is generally restricted to the aggressive arena of sports.

A second sex/touch taboo involves parents and their kids. Cuddling and stroking babies is always okay, because babies are basically asexual. As they grow up, however, contact is cut sharply as the incest taboo rears its head. Too much close contact could be seen as sexual interest, so junior isn't allowed to sleep with Mommy anymore. Girls are less restricted here again, as both parents frequently touch female children more than they do males. The mother/son incest taboo seems stronger than the father/daughter ban. The weakest prohibition is probably the mother/daughter taboo, which allows a lot of contact both early and later in life. The father/son ban is the strictest of the incest group, since it lumps the male – male rule and the incest rule together.

The intersex touch taboo is the third sex-related restriction. Here, the fear is that contact will be seen as heterosexual attraction. Any embrace between an unrelated man and woman is suspect. The prohibition even extends to couples who *are* sexually involved. Any public contact more expressive than hand-holding or a peck on the cheek will be greeted by derisive stares. The streets are off limits for most innocent intersexual contact.

There are other, specialized touch restrictions that apply to certain relationships. For example, some professions are licensed to touch. However, even these "licenses" are subject to subtle limitations. In a tough spot, a cop is allowed to get rough with a suspect, but he can't cross the line into "brutality." Doctors, of course, touch us very intimately, but they cloak this contact with a detached, professional demeanor. By handling us almost like objects, the doctor's touch loses sexual significance. Like many of the touch taboos, this can have an overkill effect, as many patients complain about the cold, clinical manner of physicians. The old-fashioned "country doctor" usually had a warm, caring touch, but it was desexualized in a different way. As a member of a small, close community, he was practically part of the family ("family doctor").

Nurses are also allowed to have a more human touch, and this is part of the role of professional comforter. Sickness makes us feel small and helpless, so we seek a surrogate-mother to "nurse" us to health. In this context, we can accept attentions which might be suspect at other times.

We regularly make contact with other professions with more limited touch licenses. Hair care, skin care, shoe and clothes fitting are all examples. In all cases, the license is severely restricted. The shoe salesman can fondle your feet, but he'd better not move up your leg. If the boundaries are overstepped, the reaction can be strong. A dentist with wandering hands can soon find himself hit with a lawsuit.

Dance floor participants also enjoy a sort of temporary touch permit. Public dancing is one of the rare occasions when the stranger touch taboo is lifted. You not only can freely tap someone on the shoulder for a dance, but you're allowed a lot of intimate contact during dancing. Religious authorities throughout the ages have sought to squelch close-contact dancing, but have never been successful for long. Perhaps people intuitively understand that dancing provides a much-needed safety valve for the energies pent up by the touch taboos.

Contact restrictions are certainly not static, since they shift over time, class and culture. For example, one recent study reports that touching between men and women has been loosening up over the last few years. Men and women have been touching each other more than in the past, a possible by-product of the sexual revolution. The other taboos, however, remain just as strong. As far as class is concerned, touch gets more restricted as you move up the social ladder. The lower classes are freest, while the upper classes are most uptight. "Primitive" cultures are also less contact-inhibited than their more "civilized" counterparts. Some societies segregate whole segments of the population from the mainstream. Such subgroups are often considered "beneath touch": to associate with them would be "contaminating" in a religious or social sense. For instance, India's lowest caste is explicitly called the "untouchables." In other cultures, a menstruating woman is kept out of contact while she remains "unclean."

In America, there are some social myths which act a lot like touch taboos. They prevent people from reaching out and touching. As you grow up physically, you're expected to also outgrow the need for certain nurturing strokes. For example, many people consider the desire to be cuddled apart from sex as childish. Similarly, psychologists often portray "back to the womb" fantasies as grossly immature. In another area, the media romanticize the image of the cold, detached success-hustler who's an "island unto himself." These myths make people guilty about all-too-human needs. Even animals appreciate the need for withdrawing into a lair to gain renewed strength. There's nothing "infantile" about a yearning for security and nurturance in a tough, uncertain world.

The isolated, "island unto himself" mentality cuts you off from mainland help during the ravages of a hurricane. Only a person with a safe, secure harbor weathers the storm. *Psychological health does require weaning oneself from one's parents, but not from the acts of parenting.* As adults, we should freely give and get parental-type nurturance from each other. The warm, loving, accepting arms of the mother should be replaced by the warm, loving, accepting arms of the lover. But sex alone can't provide all the needed nurturance. We also require lots of contact apart from sex, which is why the touch techniques described later are so useful.

CHAPTER THREE: THE TOUCH ITCH

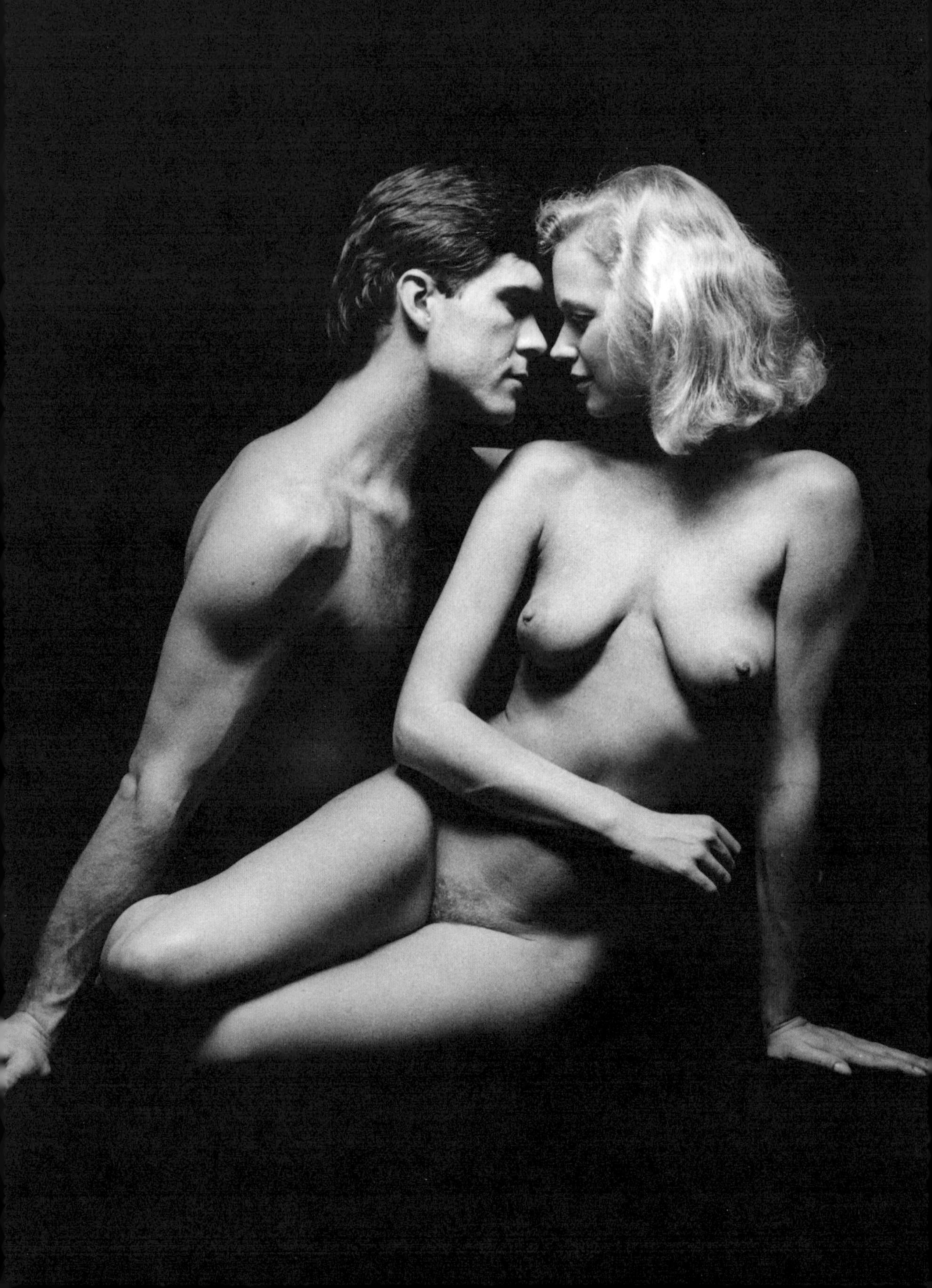

Constant, fleeting contact with untouchable strangers creates a lot of tension within us. We all wear social masks in public — a detached air of indifference which carefully hides any real interest in others. Have you ever been in a situation where a familiar face suddenly appears out of a flood of anonymous ones? This can actually be jolting, as you quickly have to strip off the mask and act truly human again.

Fact is, all people stimulate us, friends or strangers. While the social mask keeps our actions controlled, we react subconsciously to the daily encounters of a mass society. We get pushed, pressed, crowded and jostled, and all these brief brushes are like eating one potato chip or pretzel: that first nibble tempts you to down the whole bag. Unfortunately, the way things stand today, there's little likelihood of humanizing contact with strangers. There's simply too much danger, too much uncertainty. We might some day react less rigidly to the stranger touch taboo, but we'll probably never reach the stage of full spontaneity.

The antidote for the touch frustrations of daily life is not more expanded public contact, but enhanced private contact. Each one of us needs a "touch partner" as well as a "love partner." Someone has to be there to help soothe away the accumulated touch tensions. Since many advanced touch techniques are quite intimate, the most sensible choice for a touch partner *is* your lover.

While mass crowding contributes to the touch itch, it doesn't create it. We are born with the need for warm, loving contact, and the need stays with us to the grave. Tactile stimulation is a basic bodily drive, and it's surprising how often this is overlooked. This drive has been described as "skin hunger." Actually, touch deprivation has some similarities to undernourishment: both lead to stunted growth, physical and psychological. Anthropologist Ashley Montagu links early tactile stimulation with proper growth and development. In his book *Touching*, Montagu maintains that "the need for tactuality *is* a basic need, since it must be satisfied if the organism is to survive."

Touch deprivation has been connected with various disorders, including crib death, autism, asthma, and skin problems. A moment's reflection tells you that a person's basic character is deeply affected by early touch experiences. Tender, loving care tells the baby that the world can be trusted; the lack of TLC gives the opposite impression. Psychologist Alexander Lowen feels that schizophrenia may result from a lack of loving touch. The schizophrenic is often confused about the boundaries of his self. He may feel distant and disconnected from his own body. If his early body contact was cold and unfeeling, he may have detached himself from his physical sensations, seeking to find his true self elsewhere. The schizophrenic can be literally out of touch with himself.

One of the most powerful demonstrations of the touching drive is psychologist Harry Harlow's experiments with the "cloth-mother" monkeys. Harlow raised baby rhesus monkeys with two types of surrogate-mothers, one made solely with wire mesh, and the other covered with soft terry cloth. The monkeys were permitted to feed off the wire "mother," but *not* off the cloth version. Despite that, the babies still spent the overwhelming majority of their time with the cloth surrogate, clinging to the soft fabric body. When a frightening object was placed nearby, they instinctively ran over and clutched the cloth mother. Except for actual feeding, the wire mother was ignored. The experiment indicates that nourishment is not as crucial for the mother/infant bond as the mother's soft touch.

Everyone knows that human infants cry when hungry or uncomfortable. It's not so well known that crying may also signal the need for touch. Cuddling, rocking, and cooing at a little one may calm him when nothing else seems to. Unfortunately, American mothers often overlook the infant's great need for pure affection. Touching is usually done incidental to caretaking and feeding, rather than simply for its own sake. In many ways, infants are having their TLC quotients further curtailed by supposedly "modern" advances. Bottle-feeding replaces the more intimate breast-feeding (which, incidentally, is considered to be a blissful experience for both mother and baby). The old-fashiond but secure cradle is ousted in favor of the wide-open, unsettling crib.

Traditional swaddling in snug garments gives way to looser clothing, purportedly to prevent the newborn from being restricted. Home births are replaced by sterile hospital practices, which artificially separate mother and child when they need each other most. Birth is a jolting experience, and the shock should be smoothed as sensitively as possible. Warm, soothing contact is the best transition from the womb-state, and the old-fashioned procedures provided that. In contrast, the new practices fall short. Compare the cradle and crib. The cradle surrounds the infant with softness and warmth on all sides, much like the womb. On the other hand, the crib leaves the baby without such support, naked and exposed. The cold, insecure crib could be too much for some babies, which may explain the mysterious ailment known as "crib death" in which babies suddenly die for no apparent reason. Even if the infant suffers no physical harm, there could be a psychic toll. A lack of secure early contact could cause an inability to "let go." This could show up later as a fear of heights, inability to surrender to sleep, or incapacity to relax and enjoy sexual fulfillment.

Recently, there's been a lot of speculation about changes in the American character. The "me generation," "narcissistic personality," and similar labels describe a superficial type who is friendly on the surface but has trouble with lasting relationships. The narcissist pampers and indulges himself, sometimes enjoying objects which can be shut off at will more easily than people. Involvement with others actually becomes more object-like. Rather than being ends in themselves, people become means to an end: they serve the narcissist's ego, fulfill his needs, etc. When someone makes demands of the narcissist, however, he shrinks away. His weak ego is continually threatened by the possibility of being overwhelmed.

The growth of the "me generation" coincides quite nicely with the penetration of modern child-rearing practices. The post-World War II "baby boom" group is the first to have been fully raised by the new techniques, and the narcissistic character may be one result. The self-centered pampering may be the desire to get what was denied in the formative years. The object-orientation may also be a reaction to a lack of TLC. After all, if the people in your early life are frigid and unfeeling, it's not easy to become close to others later on. Harlow's monkey experiments showed that female monkeys raised with surrogate mothers developed defective maternal instincts. They had trouble with sex and treated their offsping with disdain. The attachment a baby forms with his mother becomes the model for all future relationships.

The desire for pampering is by no means the sole possible reaction to touch deprivation. The body can also substitute self-contact for needed touch. Ashley

Montagu asserts that thumb-sucking, so widespread in our culture, is absent from more free-touching "primitive" societies. The rhymthic rocking and self-hugging one commonly spies in mental institutions may be a reaction to more extreme touch starvation. The touch itch may be driven underground, but it surfaces in hidden, sometimes destructive ways. After the toddler years, the main source of touch for many children comes through discipline. Some kids misbehave, consciously or otherwise, to court the bitter-sweet contact of a spanking. For some, painful contact may be better than none at all. This can warp the psyche, creating an unnatural association between pleasure and pain, the roots of sado-masochism.

Healthy or not, aggressive contact is the main touch outlet for many adults, particularly men. Compared to females, the American male really gets the short end of the touch stick. For one thing, men are subject to much more stringent touch taboos. For another, the basic male role model is the antithesis of tender touch. Coolness, detachment, competition, manipulation, dominance, toughness, shrewdness, and similar traditional male traits are hardly the stuff of which TLC is made. With this in mind, the great popularity of contact sports among men is hardly surprising. (The growing interest in women's contact sports coincides with the increasingly male-like roles played by women in many areas.) Rough contact sports allow intimate touch cloaked in the proper macho trappings.

It's interesting that some of the most spontaneous and close male-to-male contact comes during sports *apart* from the actual play. Football, hockey and soccer players burst into wild, uninhibited embraces after scoring key goals. In any other context, this intimacy would threaten male identities. Here, however, the sports activity affirms the male role and permits otherwise prohibited touching. The sports scene also caters to touch-starved spectators. The taboos are relaxed in the stands. Key plays permit spontaneous, joyous contact. The fans also seem to enjoy vicarious thrills from the contact on the field. The more strenuous the contact, the more emotionally gripped are the fans. Tennis and golf attract more subdued followers, while boxing, hockey and football draw more enthusiastic throngs. Professional wrestling is an oddity: the matches are usually faked, the rough-stuff is staged, and the atmosphere is more carnival than competitive. Yet all over the world, pro wrestling brings out magnetized, frenzied fans. Whatever else can be said about wrestling, the intense skin contact between the combatants is obvious.

Unfortunately, sports is not the sole form of aggressive touch. *There are several links between violence and touch deprivation.* We have seen how discipline as the main childhood contact can lead to sado-masochistic tendencies. Lack of TLC also lessens our ability to empathize with others, which makes it emotionally much easier to inflict pain without guilt. Finally, an untouched adult has lost a valuable safety valve for his tensions. He is not soothed and healed by calming touch, so he becomes more prone to violent outbursts. His tolerance for life's usual frustrations is diminished.

People also find touch-substitutes apart from aggression. Some of the most touch-inhibited nations have the largest number of pets per person. Like babies, few touch taboos apply to animals. No one will look twice if you stroke, cuddle or caress a neighbor's dog, or even a stray. Because of our fear of old age, senior citizens are even more touch-deprived than most, and they frequently use pets to fill the contact-void. Pets become a "safe" touch-outlet even for those who are excessively touch-neurotic with people.

The well-known American sex obsession may also be related to touch. Sex still remains the prime path for satisfying (though indirectly) our deep-seated touch itch. The invitation to the bedroom is also an implicit invitation to embrace. This gives sex all the more power, and may help explain why sex is used to sell just about everything in the U.S. The appeal may actually be to two basic drives, sex and touch.

Repressed touch needs may also be present in the American obsession with bathing, both sun and water varieties. The beach can be a real skin orgy. We're blissfully caressed by the ocean's embrace, gently stroked by sea breezes, and warmed by the sun's soothing rays. At home, the bath relaxes and refreshes us, and the shower tingles and brightens us. Perhaps that's why the tub prompts so many people to spontaneous song.

Another common compulsion, oral cravings, may also be touch-connected. The lips are among the most sensitive touch-receptors. For babies, the breast, bottle, and even the foodless pacifier all act to calm him. In later life, tensions can also be defused by sticking something in the mouth. Thumb-sucking in childhood gives way to nail-chomping, cigarettes, overeating and overdrinking. The last two may result from the fact that food and drink going down the digestive tract provide a sort of internal caress, leading to a sated, secure feeling inside. Problem is, these tension reducers, effective though they might be, become harmful addictions. *Cutting tension directly through soothing touch techniques is a far healthier alternative.*

The everyday workings of the touch itch can be seen anytime in local stores. Store owners know that it's almost impossible to keep customers from fingering items on the shelves. One toilet tissue manufacturer capitalizes on this urge in its ads, which show a flustered manager futilely trying to keep shoppers from squeezing the product. Shoppers somehow feel that touching something is the ultimate test of its quality. Our eyes give us a general picture, but to *really* know something, we seem to want to lay hands on it.

The widespread presence of cuddly toys in stores is another recognition of the touch itch. Though among the most popular gifts, they are found in many stores other than gift shops. Merchants recognize the strong attraction we feel for soft, stuffed animals, which are even some of the most sought-after prizes at carnivals and street festivals. A common sight at one of these events is a teenage couple with the girl clutching a stuffed teddy which her boyfriend won for her.

Of course, there are more direct routes to sate our touching drive. For many, touch professionals like beauty parlor operators or masseurs help scratch the itch. Still, the repression remains, as touch is frequently hidden as even a secondary motivator. We go to the professionals, we say, for health and beauty reasons, not to enjoy "sinful" touch. Whatever the rationalization, the reality is that touch *is* a very basic need. And, while it's crucial to early development, the need doesn't cease with childhood.

Tactile stimulation enlivens our bodies, sparks our senses, heals our hurts. In many respects, being touched gently means being loved. As I've tried to show, we can repress our conscious desire for contact, but our body still feels the ache. If we ignore it, the itch will find substitute scratching-posts, some of which may be harmful. A healthier attitude is to actively and joyously cater to your skin hunger.

The message of the last chapter was simple: our touch instincts could use a much larger dose of direct touch. There are different types of touch, but much of what we daily receive doesn't satisfy our deepest desires. Furthermore, fleeting and anonymous contact actually enflames our tactile appetite. It's a bit like passing a bakery and sniffing the wonderful aromas—our juices start flowing and we crave a taste of the real thing. These brief brushes are like touch foreplay—they arouse the instincts without fulfilling them.

The same is true of the common contacts we make with friends and acquaintances. We *do* touch, but the contact is so superficial and ritualized that it loses much of its ability to sate the itch. Handshakes, pats on the back, pecks on the cheek and similar greetings all show affection, and life would certainly be starker without such tactile amenities. However, these gestures are only "skin-deep"— they stroke the surface and no more. But our biology demands more. Soft, surface strokes are a preliminary type of tactile delight. We also need deep pressure for complete touch fulfillment. Only tight, snug contact touches us to the core.

In ordinary life, deep touch essentially comes in one form: the full embrace. And we all know how restricted this contact can be. It's generally reserved for close relatives, bosom buddies, and lovers. Even with them, you mainly see full embraces on such special occasions as reunions and farewells. And on those rare instances, it's usually inhibited. The most common form of the full embrace has the pair leaning forward so the chests touch, but the lower bodies remain rigidly apart. This is to avoid genital contact and its inevitable sexual link. Once again, the taboos emasculate the touch experience.

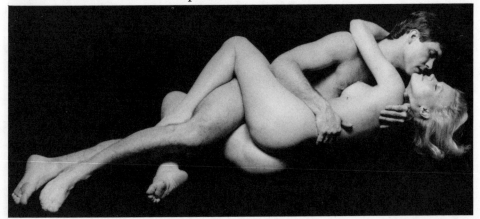

Even if we were somehow catapulted into an ideal contact culture where free, full embraces were the norm, we still wouldn't reach the inner core of the touch itch. We need that full embrace *plus* something more: to be lying down. When lying on a comfortable surface like a bed, more of us is being snuggled. An added bonus is support, which makes us feel relaxed and secure. The bedroom embrace also permits a lot more closeness than the standing version. Your lower bodies can be cozily wrapped together for the fullest possible contact. What I'm describing, of course, is cuddling, although it's also been called snuggling, nesting, snoozling, and heaping. Whatever the label, we have a positive craving for it.

The cuddle crave dates back to our time in the womb. Because of that, it's the most basic part of the touching drive. It's hard to overestimate the power of such a deep-seated memory as being engulfed within the mother's body. But the cuddle crave is not solely built on the womb-state. It's reinforced by our initial years of warm, maternal embraces. Our tiny frames continued to be lovingly enveloped by the mother's larger figure. And as we grew, the cuddles continued, often at the hands of others besides the mother. The touch taboos cut this off eventually, of course, but not before plenty of strong conditioning has occurred. Psychologists say that early childhood experiences are crucial for later life. We're continually exposed to various forms of cuddling from conception through our first few years, a potent influence indeed.

Evidence exists that a proper dose of childhood cuddling is needed for sound psychic development. When Mother can't be around, that teddy-bear or Linus-blanket can be a healthy substitute. A report in the *Journal of Child Psychology and Psychiatry* (noted in *Psychology Today*, Jan. 1980) found that children who habitually snuggled soft objects were more independent, went to bed more cheerfully, and slept better. Emotionally-troubled youngsters tended not to use a cuddly object. The rate of emotional disturbance was almost twice as high among the non-cuddlers as among the cuddlers. Of course, most of us outgrow old teddy. But what about flesh-and-blood cuddling? Just because kids need it, does that mean adults necessarily don't?

Available evidence strongly suggests that the cuddle crave endures into adult life. Psychological studies show that females feel a strong urge to be held. In a report by psychiatrist Marc Hollender, one woman described the craving as "an ache . . . a physical feeling." In *The Hite Report* on female sexuality, many informants spoke about the need to be held. One of them put it like this: "The embrace, which involves the whole body, is important to me. Having my naked body lying against the naked body of my partner — especially my full front touching my partner's full front."

The studies also reveal that some women need cuddling so badly they swap sex for it. Some observers have speculated that female promiscuity may not be so much the result of hyper-sexuality, but a strong cuddling urge. The same studies show that men see women's desire to be held as the first move towards sex. They apparently don't appreciate the cuddle need as an independent drive. The likely reason for such shortsightedness is the poverty of touch enjoyment allowed men. As already emphasized, the touch taboos are much more restrictive for men. Females are stroked more frequently from infancy on, so it makes sense that they'd be more in tune with their needs. On the other hand, male touch drives are driven further underground by the widespread macho role models. Cuddling is popularly connected with childishness and dependency, and it's threatening for the average man to associate himself with such traits. Men need to understand that it's natural to relax and be nurtured in the arms of another. As men learn the benefits of touch techniques, they'll start singing a different tune.

Like the touching drive in general, a repressed cuddle crave works its way into other forms of action. The most common cuddle-substitute is sex. Zoologist Desmond Morris did an interesting survey of the intercourse postures pictured in illustrated sex manuals. He found that 74% of these positions contained some act of embracing which had nothing to do with steadying the body for sex. "All these actions are basically embraces, partial embraces, or fragments of embraces," he says in his book, *Intimate Behaviour*. "They indicate that, for the human animal, copulation consists of the adult primate mating act *plus* the returned infantile embracing act. The latter pervades the whole sexual sequence from its earliest courtship stages right through to its final moments."

It's possible that the cuddle crave is intimately connected with the sex drive. Anthropologist Ashley Montagu has described the sex impulse as divisible into two parts. One is the aspect we normally associate with sex, genital satisfaction. The less obvious component is the need to make close contact with the partner's skin. In a touch-healthy individual — one who has received adquate TLC in the formative years — both aspects will be intermingled in sex. A lack of childhood contact, however, could throw the two out of balance.

The cuddling crave can itself be divided into two parts. One is the need to be held, and the other is the desire to hold. The pleasures of being held trace directly from the womb-imprinting and childhood TLC. We seek to recapture the comfort, security, warmth, peace and undemanding love we felt in our mother's loving embrace. This comforting contact is doubly important in our frenzied, insecure world, where nothing seems nailed down and the only unchanging reality is change itself. Far from being immature, the life-sustaining nurturance of a lover's embrace is a completely appropriate way of coping with the demands of a harried existence. When something frightens a child, it instinctively retreats to the security of the parental embrace. Renewed, the child then feels encouraged to go out and explore again.

While adults can't retreat to a lover's arms after each joust with life, they can gain a similar renewal by being held each night. The snug embrace can be a safe harbor in troubled times.

The other component of the cuddle crave — the desire to hold and comfort another — is part of our parental instincts. We react this way to children, small, helpless creatures, and to anyone who's in a child-like state. For instance, a sick, injured or aggrieved adult may evoke this comforting response even from strangers. In love-play, men often respond to women in this way. The female's relative fragility, smallness, and delicateness prompts the male into protectiveness. He is moved to cherish her like a child. Actually, though, there's a child in each of us desiring to be cherished and snuggled. There's nothing infantile about surrendering to such impulses, as long as they don't dominate your emotional life. What *is* infantile is the effort to remain isolated and aloof, cut off from your cuddle craves.

Touch-partners should alternate between holding and being held. That way, the full cuddle crave is completely sated. Regular, sustained cuddling between two people leads to a special type of intimacy I call a "cuddle bond." When you cater to each other's cuddle needs, your bodies never seem to be fully separate, even when you're physically apart. It's as though you incorporate the other person's body into your own.

Cuddling builds a special state of trust and acceptance, a fact recognized by modern sex therapists. A common technique for treating sex disorders with couples is to have them cuddle together without sex. The cuddling gradually produces the confidence and security needed to rebuild the sex bridge. Cuddling soothes tensions, relaxes uptightness, and lets lovers be open and uninhibited with sexual exploration.

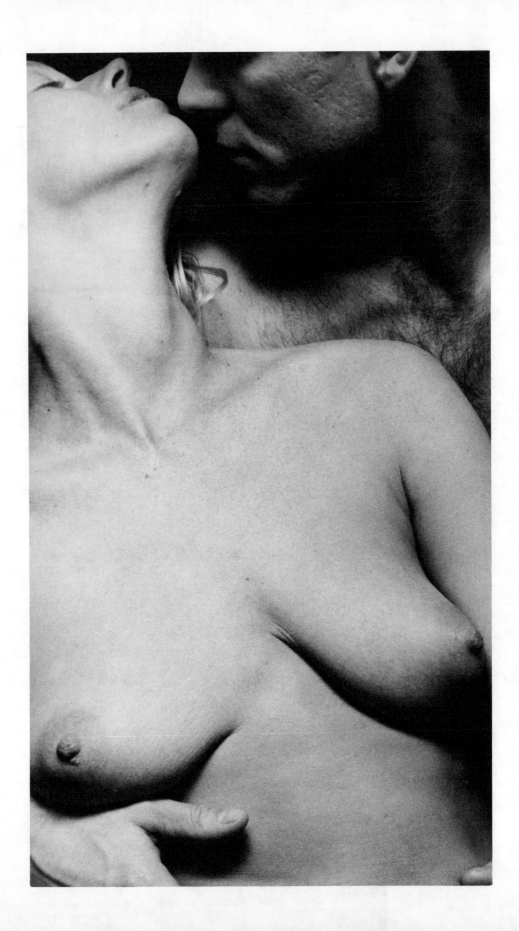

PART II: TOUCHING TOOLS OF THE BODY

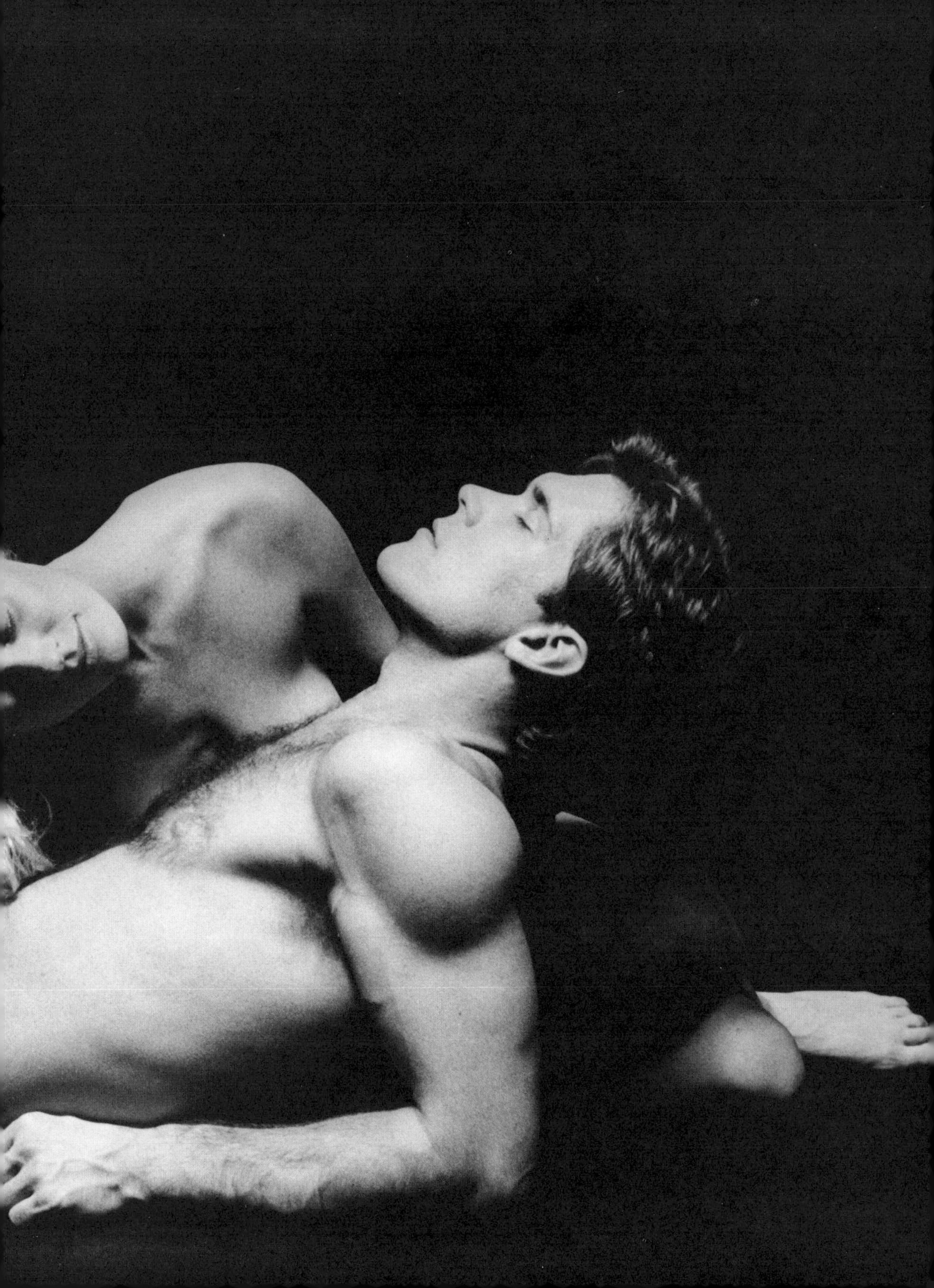

CHAPTER FIVE: THE SKIN

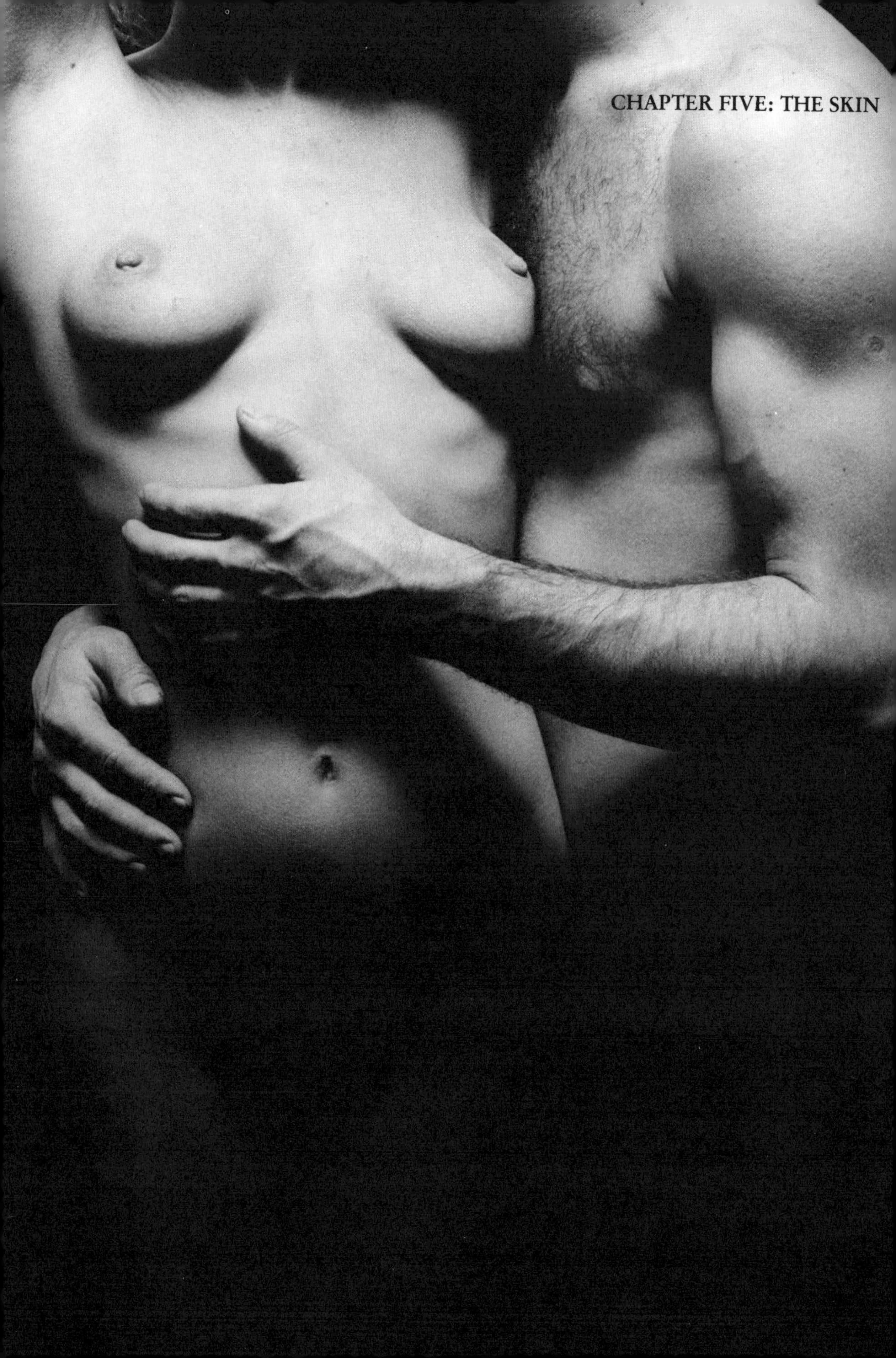

The skin is the play upon which the touch script is written. Man stands virtually alone among primates in possessing such a large, smooth surface for tactile strokes. Unlike our furry cousins, our relatively hairless bodies erect few barriers for direct stimulation. It's almost as if we were built for maximum touch arousal.

But the skin's tasks travel far beyond touch. Perhaps because it's constantly before our eyes, we overlook the crucial roles played by the skin in our lives. The skin is actually the largest organ system of the body, covering about eighteen square feet, and weighing about eight pounds. It's also the first sensory organ to function, since tactile sensitivity starts in the earliest stages of womb-life. The other four senses — sight, hearing, smell and taste — are really specialized offshoots of the embryonic tissue which produces the skin. This tissue, the ectoderm, also gives rise to the central nervous system (CNS). Because of this common origin, Montagu and Matson described the skin as the "external nervous system" (*The Human Connection*). Certainly, the CNS and the skin are close partners, one regulating the body's internal states and the other controlling its relationships with the external environment.

The importance of the skin is reflected in the large tactile area of the brain. Compared with other senses, tactile functions are proportionately over-represented in the cerebral cortex. Moreover, the nerve fibers which transmit touch sensations are usually bigger than other sense nerves. And the skin is rich in nerve fibers, not only for recording touch, but for temperature, pressure and pain as well. These sense receptors constantly send information to the brain concerning our connections with the outside world. *If the brain can be compared to a computer, the skin can be likened to the operator who feeds data into it.*

Actually, the skin's function is hardly limited to communications. It also protects the internal organs from injury and disease, and regulates body temperature through heat loss and perspiration. The skin contains thousands of specialized glands: sweat glands control perspiration; oil glands moisten the hair and skin; mammary glands (in women) secrete milk. The outer layer of skin, the epidermis, is relatively thin and free of blood vessels. The inner, thicker, layer — the dermis — is richly supplied with blood and lymph vessels, nerves, glands, hair follicles and fat cells. Skin thickness varies throughout the body, with some spots (like the soles of the feet and palms of the hands) being much thicker than others. The number of hair follicles also varies, but the palms and soles are the only completely hair-free body parts. Human hair and nails — like animal fur, feathers, scales, claws, hoofs and horns — are also skin derivatives.

Another interesting skin quality is its ability to conduct electricity, which literally passes between people when they touch. "There can be little doubt that in tactile stimulation electrical charges are transmitted from one individual to another," says Ashley Montagu in *Touching*. The skin conducts electricity quite efficiently, which is why lie detectors are able to take readings from the skin. Strong emotions amplify the electrical conductivity of the hands and feet. Emotions can also cause the blood vessels in the skin to expand or contract. When someone is anxious or fearful, he literally gets "cold feet." In contrast, two lovers together may find their skin flushed from increased blood flow.

The skin is sensitive in several other ways. We can identify different objects solely by touching them. We also learn to differentiate people by the way they touch us. Dr. James Gibson of Cornell University demonstrated that people could detect movements across the skin as slow as a millimeter a second — the same speed as the minute hand of a large clock. The skin senses the location of distances on its surface more accurately than the ear can read sound distances.

Ashley Montagu has noted some interesting connections between the skin and emotional conditions. People who suffer sexual abnormalities frequently experience skin disorders as well. The common link may be a lack of early TLC. Those afflicted with sex or skin problems often receive inadequate tactile stimulation as infants. One study of 25 eczema sufferers showed that the majority had been deprived of maternal skin contact. It's as though the skin were writing the message of its early deprivation on its own surface. The skin can also reflect less severe conditions, like ordinary stress. Common reactions to stress include scratching the hair, rubbing the face, pulling the ears, etc. Such skin strokes serve to soothe our tense nerves and placate our busy hands.

In the realm of touch sensitivity, not all our body parts are created equal. A 19th century German scientist, Ernst Heinrich Weber, performed a series of experiments which proved that "tactile acuity" (his term) varies in different areas of the body. Weber carried out his tests with an adjustable metal compass, the kind used for drawing circles. He varied the width between the compass legs until he found the smallest distance each body could detect. The most acute touch-spot turned out to be the tongue-tip, followed by the fingertips and the tip of the nose. The least accurate touch-spots included the spine, mid-upper arm and mid-thigh. Falling in the middle were places like the buttocks, knees, neck and shoulders. Generally, the face, palms of the hands and soles of the feet have the most sensitive touch centers. Areas on the trunk are generally less sensitive than the head or limbs. Usually, the ends or projecting parts of any place are more sensitive than the middle, internal areas. Muscular spots are basically less touch-accurate than other parts.

Weber also found that the left side is more sensitive to touch than the right side. He explained the shifting touch-responsiveness throughout the body to the density, pathways, and endings of nerves, which differ in various spots. However, he emphasized that the "tactile acuity" he was measuring is *not* the same as other touch sensitivities. In the English translation of his book, *The Sense of Touch*, he says that "even the breasts, whose nipples are very refined, and which are considered to have good sensitivity, lack an acute sense of touch. Those people who confuse tactile acuity with the ability to receive a certain titillation or a sharp pain are very much mistaken."

The breasts, like other erotic areas, are built for a different type of tactile sensation than Weber studied. The erogenous zones contain a special erectile tissue which becomes engorged with blood during sexual excitement. The blood helps to harden the area, making it more accessible to sexual stimulation. In the case of the penis, of course, blood engorgement provides the erection needed for penetration.

A second type of specialized touch response is ticklishness. Psychologist Havelock Ellis has speculated that ticklishness may have started as a warning system, since many prime tickle-spots like the thighs, palms, and soles of the feet are particularly vulnerable. Therapist Eleanor Hamilton says that tickle spots represent pleasure areas in tension. Whatever the explanation, ticklishness is often closely connected with sexual reactions, since many tickle and erotic spots overlap. *In the chapter on tickle play, we shall see how tickle techniques can enhance both your sex and tactile repertoires.*

Apart from different touch sensitivities, body areas can also be classified as *active* or *passive* touching tools. Passive parts like the tummy, thighs, buttocks and face enjoy receiving touch sensations, but are too immobile to return them. Active areas like the hands, tongue, lips and nose are equally adept at giving and getting tactile delight.

CHAPTER SIX: THE HANDS

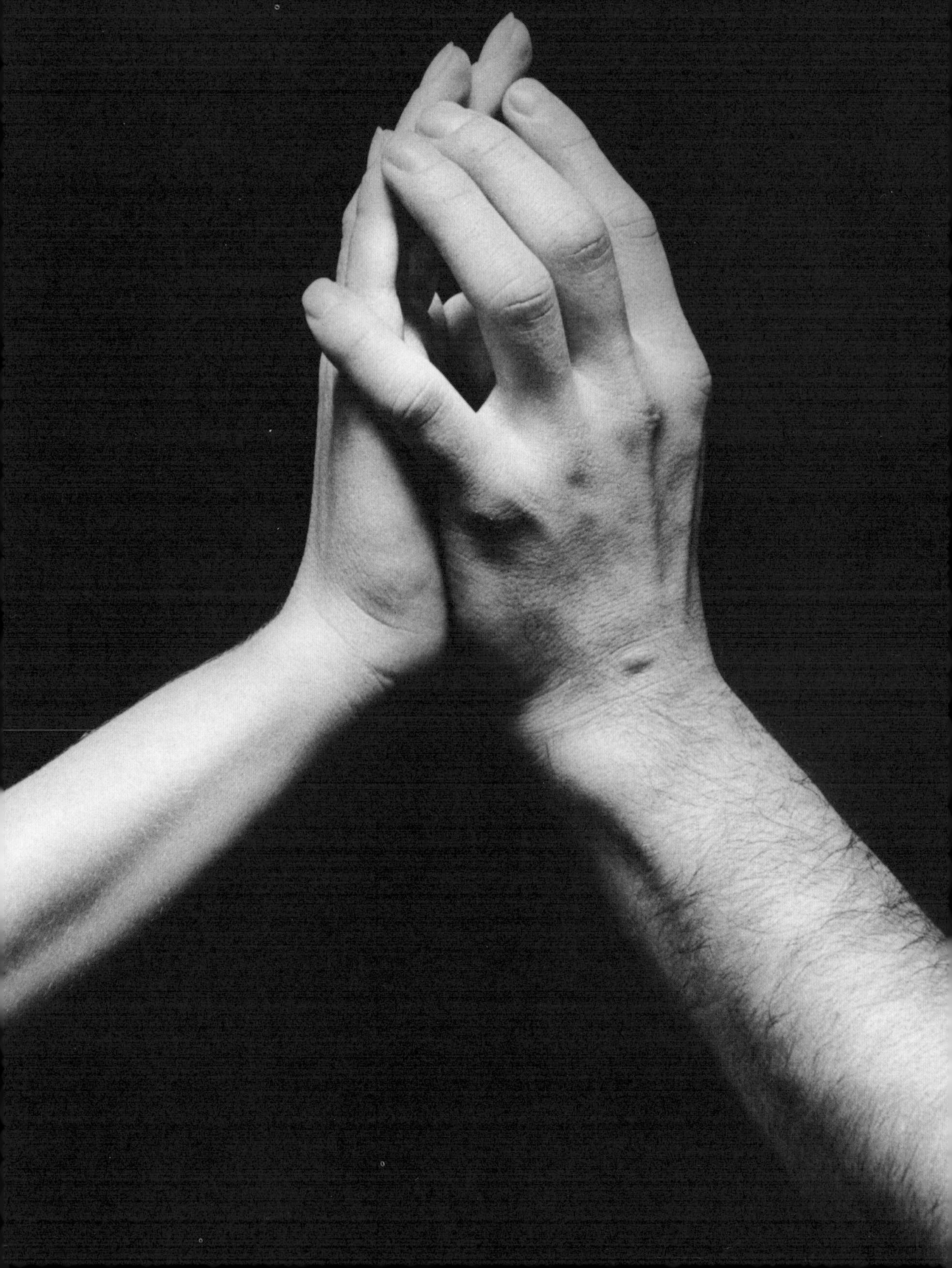

The hands are easily the most versatile touching tools of the body. The fingers, thumbs, palms, knuckles and fists offer countless combinations of pressure, strength and texture. Your imagination is the only limit to the touch "handiwork" turned out by your hands.

Actually, it's hard to separate the concept of touching and the hands. There are few forms of contact in which the hands don't come into play. Even when they aren't the main tactile focus, they still play significant supporting parts. For example, sex and cuddling would be pretty tough without the balance provided by the hands. They are essential props in virtually all touch scenes.

In many respects, the hands are a real mark of our humanness. Man took a great leap forward when he developed fully opposable fingers. Unlike animal paws, our fingers and thumb can stand completely opposed to each other, making precision work possible. Without this ability, our tool-making capacity would never have reached its high development. So in a real sense, we owe our civilization to our hands.

Like the words "feel" and "touch," our culture has plenty of metaphors which reflect the true importance of the hands. For example, a "handy man" is "dextrous" at many different tasks. A "handicapped" person is restricted in some sense from normal "manipulations." Our sophisticated "digital" computers "hand" us good data. Bums seek "handouts," which they often get from a "handful" of pocket change.

The significance of the hands is also reflected in their universal use in greetings. Offering one's hand for a handshake is a common sign of friendly intent. Waving the hands serves to greet someone at a distance. The hands are also used in less flattering communications. An upturned middle finger rather straightforwardly tells someone where to "stick it." In some cultures, hand gestures tell as much of someone's story as words do. Sign language for the deaf is simply an elaborate extension of the use of hands as communications tools. Hands also say a lot through touch. A person's mood, attitude, and basic temperament are all reflected in his fingertips. A "cold fish" handshake will make you cringe, while a warm, friendly clasp will prompt you to open up.

The hands can be considered the handmaidens of the mind. No other organ is equipped to carry out our wishes the way the hands can. They help you eat, drink, go to the bathroom, and have sex. They clean and care for the rest of the body, and are on call whenever you need them. What animals accomplish with their tongues, we do with our hands. They groom, massage, and give us pleasure. They also cater to our unconscious needs. Many of our compulsions involve the hands—nail-biting, thumb-sucking, smoking, hair-stroking, obsessive hand-washing. The hands and face work closely together to reduce stress. When tense, we instinctively rub our noses, scratch our heads, play with our mouths and chins. A sort of self-mothering, comforting response.

To make the most of the hand's touch potential, we use all of its component parts. Different touch techniques require a variety of pressures, and the hand is well-equipped for this. The skin responds well to two types of touch. One is an exquisitely light, gentle stroke, the feather—touch or tickle-stroke. This turns on the skin surface, causing a spine-chilling, tingly sensation. The finger tips are the prime performers here, since they're capable of the lightest, softest contact. If you want to feather – touch with a larger surface, the flat part of the fingers and the palms can also do the job.

A second type of satisfying touch is heavy and deeper, and is useful in massage and tight embraces. This causes a soothing pressure which loosens and relaxes stiff muscles. The thumbs, palms, knuckles and fists (and sometimes, even the elbows) are all useful here. You can increase the pressure from these spots by applying body weight.

Whichever touch-type you use, your hands should be smooth and flexible. Calloused, hard hand surfaces are a real turn-off for your touch partner. If you do a lot of manual labor, try using hand-care products regularly. A related problem is being "all thumbs." Some of us are born with large, relatively immobile hands. Others are clumsy because they rarely do close work. But you must be a "soft touch" with your touch partner. That's doubly true if you're a hulking, muscular, macho man. Your toughness has no place in touch play. You don't want to poke at her as though she were some sort of machine tool. If you're burdened with rigid, uncoordinated paws, try practicing over and over some close, careful task like threading a needle.

The hands are also significant in certain systems of esoteric philosophy. The ancient practice of palmistry purports to predict your destiny and fortune by reading lines in your palms. Other systems associate personality types with different hand sizes and shapes. For example, the square, broad hand was supposed to mean an earthy, practical person. The thin, rectangular hand hinted at a dreamy, ethereal disposition.

In certain massage systems, the energy is said to flow along invisible pathways in the body called "meridians." Many of these meridians start or end at the fingers or toes. One system using the meridian concept is called reflexology. Points in the hands, feet and other body parts are supposed to be reflexes to different glands and organs. For example, pressing and massaging the thumb benefits the head. A thorough hand massage will stimulate the entire body. Hand massage techniques will be described later in the sections on massage and health techniques.

The hands have always had a central role in healing. "Laying on of the hands" appears again and again throughout history as a method for alleviating illness. Psychic healers and practitioners of traditional Oriental medicine agree that life energies emanate from the fingertips and can be transferred from one person to another. This energy transfer will be described in detail in the section on healing.

So far we've stressed the hand's role as a touch-giver. However, it's equally adept at receiving touch, since it's one of the most contact-sensitive spots on the body. Hand massage is one way to give hands the contact they crave. Another way is to merge the giving and getting roles through hand-holding (or, to put it another way, hand-cuddling).

In this age of sexual liberation, big thrills like intercourse and orgasm get lots of attention, while subtler pleasures are often neglected. Hand-holding between lovers has been overlooked lately, which is too bad. It's one of the best forms of touch-sharing, since you're mutually pressing highly sensitive tactile-spots together. You can also do it anytime, even in public. In simpler times, the Beatles sold millions of copies of their hit record, "I Wanna Hold Your Hand." Their tip is still worth taking today.

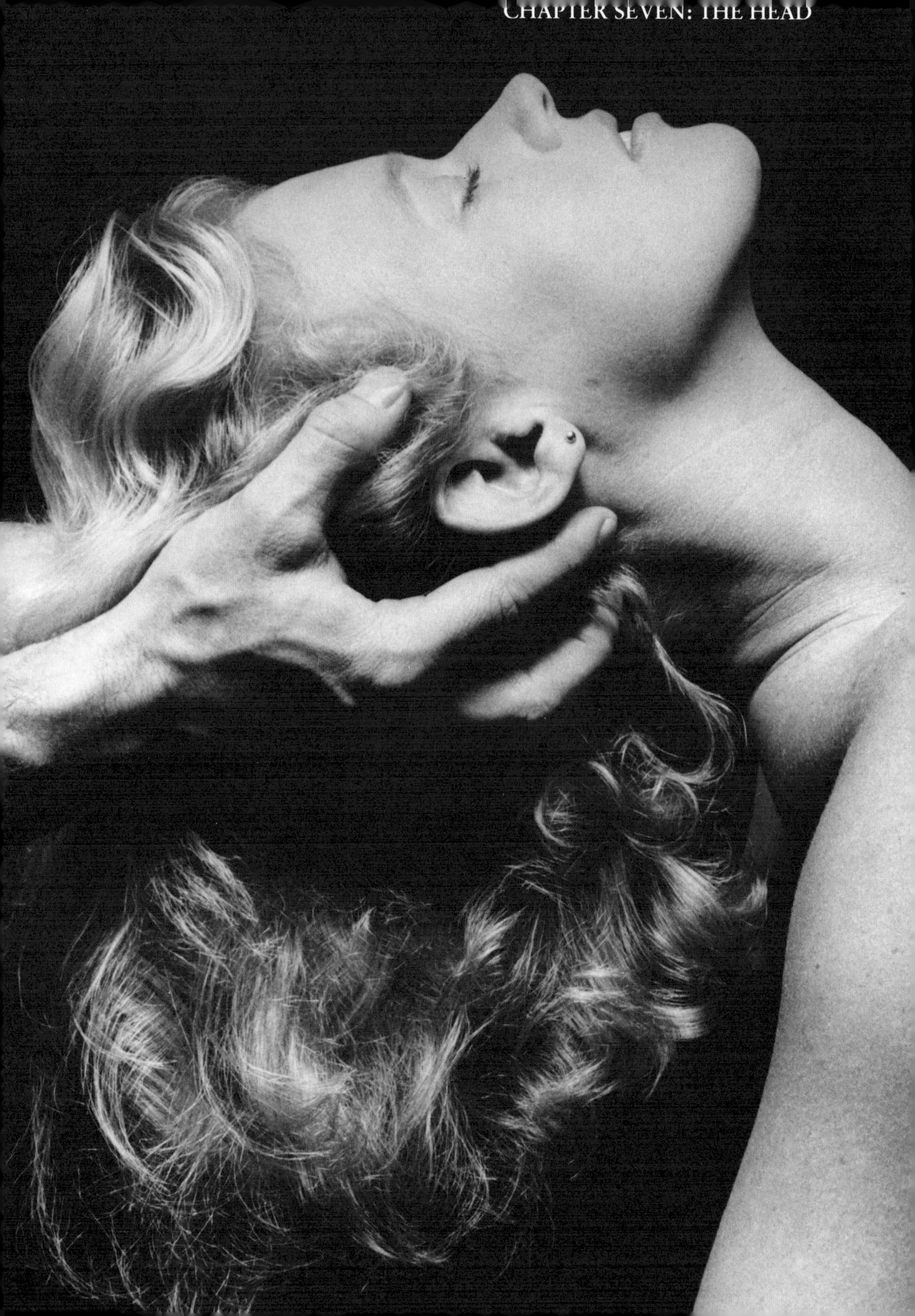

The head is a marvelously varied instrument of touch. It offers an endless diversity of textures and terrains. The soft, round surface of the cheeks contrasts with the thin skin surrounding the forehead and jaw. The peninsulas of the nose, ears and chin stand out amidst the flat plain of the forehead and the valleys of the mouth and eyes. The hair on the head is denser than it is anywhere else on the body. The head is also unique in housing five separate body orifices, divided among the nose, ears and mouth.

Despite this wonderful complexity, the head is pretty much neglected as a touching zone. Few pleasures in life rival the experience of a loving hand on one's face, but this happens fairly rarely. Although the head is one of the most accessible skin areas — it's never fully covered by clothes — it remains one of the most private. You can shake a buddy's hand, poke his shoulder, slap his back — but don't go patting his head. If someone does that, it's usually taken as a condescending put-down on the part of the patter. We consider head-strokes a sign of parental fussing, so we're leery about receiving them as adults. Mother always seemed to be pawing your head — testing your temples for temperature, poking around your face with a washcloth, combing your hair, wiping the corners of your mouth with a napkin.

These early childhood experiences — some of which were annoying and even humiliating — may make us overly-sensitive about head-strokes. Another reason may be the vulnerability of many areas of the head. We instinctively pull back from anything poking at our eyes. Our noses are easily broken, our skulls easily cracked. We fear facial injuries more than any others. Therefore, facial intimacies are pretty much restricted to people we're particularly close to. Lovers can and do run fingers through each other's hair, pat faces, and hold heads in each other's hands. But even with lovers, head strokes are not too common. A rediscovery of facial delights is long overdue.

Head touching has an added dividend for the rest of the body. Nervous tension collects in the muscle groups on and around the head, so, working these areas can have soothing effects all over. If you find yourself short on time for a full body massage, a facial massage can be a shortcut to relaxation. (See the massage section for tips on facial massage.) For good tension relief, try finger-tip pressure over the eyebrows and down the bone surrounding the eye sockets.

Fingertip pressure over the eyebrows   *Scalp rub*

Also try gently pinching the bridge of the nose. These spots contain numerous "acupressure" points which trigger relaxation reflexes. Other tension pockets needing work are the forehead, the scalp, and the muscles which start at the base of the skull and go down the neck. Knead, press and rub all these spots for a rejuvenating response. The scalp skin is especially in need of soothing stimulation. It's hard for this area to get enough air, particularly with the junk people put in their hair. A row of acupressure points runs down the center of the head, starting at the forehead and going down to the neck. A scalp rub using the fingertips or knuckles can be really exhilarating. Gentle hair pulls all over the head also wake up the scalp. Other techniques for invigorating the area — like head-tapping and slapping — are discussed in the section on sensitivity strokes.

Some spots on your head really crave gentle touching. The cheeks and the area around and under the chin thrive under soft, fingertip feather touches. A piece of fur, silk scarf, or ostrich feather lightly moved over this area offers similar titillation. This is spine-chilling, goose-bump stuff. A more soothing effect comes from rubbing with the palms or cupping the face tenderly in two hands. Gentle palm pressure over closed eyes is also relaxing and restful. The ears are fertile ground for touch-play. Light finger strokes over, under and inside the ears produce an orgy of tactile thrills. Ear pleasures easily become pure eroticism when you bring the lips and the tongue into the fray. Who can remain deaf to nips of the lobe and outer ear, or a tingly wet tongue-tip probing the inner parts? Gentle blowing into the ear provides a finishing touch to this "ear-oticism."

The mouth, with its mobile tongue and lips, rival the hand as the body's top touching tool. The mouth rates highly both as a touch-giver and touch-receiver, since the lips and tongue are quite contact sensitive. Such a simple act as lightly running your fingertips over the lips provides sweet proof of their tactile responsiveness. As a touch-giver, the mouth offers something no other spot can match — a moist lubricant which is naturaly applied while touching. This adds an immeasurable dimension to your touch-play. For example, a gentle fingertip massage can stimulate and invigorate the entire body. You can enhance this massage with a light licking action with the tongue and the lips, but

Lip play   *Ear Pleasure*

Mouth play   *Hair strokes*

you've moved into a whole new ballpark then. The mouth massage turns simple sensuality into eroticism, since the association between mouth-play, saliva and sex is so strong. It's pretty hard not to be turned on by wet, tongue-kisses, and the mouth massage applies these all over the body.

The mouth is more than just a touching tool. It's also a talented erotic instrument. We'll get into the subject of the mouth a bit more deeply in the section on erogenous zones in the next chapter.

As already mentioned, the hair-covered spots on the head need stroking and stimulation. This includes not only the scalp, but the eyebrows and facial hair (for men). In two situations, hair can also serve as a touch-giver. If you have long, flowing hair, you can give your partner an exquisite tactile high by running your locks all over his nude frame. Make slow, tantalizing sweeps across the front and back. This can be done in conjunction with a massage, or just for its own sake. Men also get a lot of pleasure from nuzzling up behind a female possessing long locks. A woman can receive a similar sense delight by snuggling up to a hairy male chest.

The second active hair technique involves the eyelashes. They can be used to supplement tickling strokes and fingertip feather touching. For example, get close enough to your partner's face so your eyelids are barely in contact with his skin. Then blink in a steady rhythm as you slowly move your eyes across his face. The tiny probing hairs produce one of the softest touches possible.

Before leaving this subject, some mention should be made of the role of the eyes in the touch experience. Physically, the eyes can neither give nor get touch, although the closed eyelids can be lightly stoked. The eyes *do* touch us, but in a very unique way. Eye "contact" occurs when your gaze directly meets someone else's. The stare, even if only fleeting, "touches" us deep within. It can make us feel naked, trapped, exposed. It's noteworthy that staring among strangers is governed by prohibitions which parallel the touch taboos. Strangers are restricted from touching with either bodies or eyes.

Some people can't look *anyone* in the eye because it threatens to tear away the social masks. Instinctively, they fear that their eyes will reveal the hidden truth about themselves. Since avoiding eye contact is so common, it's useful to lock eyes with your touch partner once in a while. Just stare at each other silently for a few minutes, letting your orbs carry out their conversation. This won't be easy for some, because the sustained stare can make you blush, flinch or turn your eyes away. But it can lead to greater communication and trust. After trying it, talk about your experience with each other. Compare notes on the non-verbal intuitions you both received during your prolonged gaze.

CHAPTER EIGHT: THE NOSE

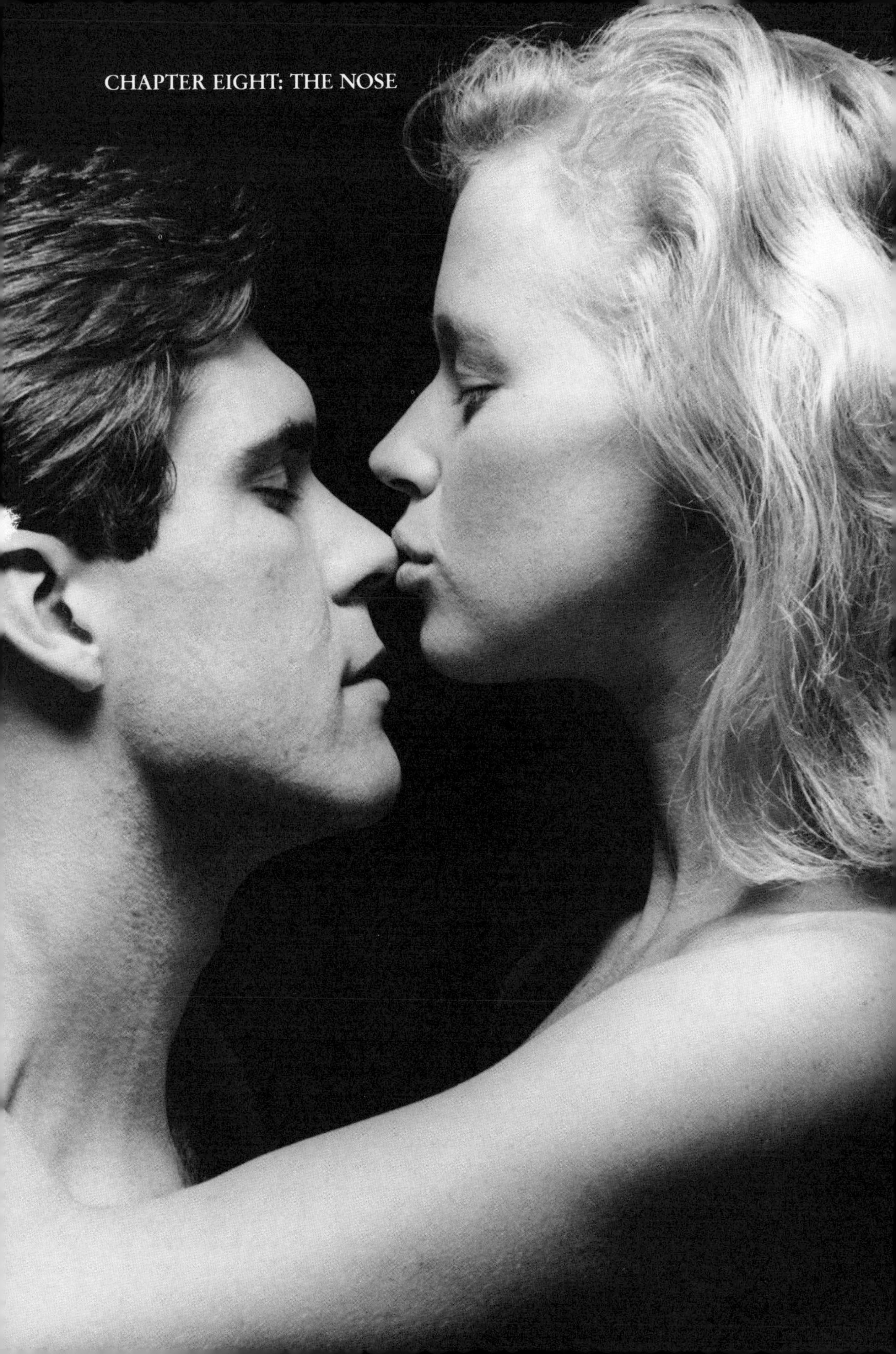

I've devoted a separate section to the nose because it's such an intriguing, but grossly ignored, organ. It is amazing how we overlook the potential of the nose. We all know about the nose's role in smell, but that's just a small part of its job. For one thing, the nose makes a direct connection with the emotions. The olfactory (smell) nerves are linked with the limbic system in the brain, which deals with instincts and emotional states. The simple movement of air through the nose sets off electrical activity in the olfactory nerve and limbic system. Breathing and smelling are more important for the way you feel than you may have suspected. Poor, irregular breathing may be associated with poor, irregular feeling. Certainly, perfume manufacturers are aware of the link between the nose and the emotions.

This may also help explain why touching the nose calms many people. When someone is tense or nervous, the nose often gets rubbed, scratched, tweaked, and picked. These movements have a soothing, mellowing effect. It turns out that the limbic system contains the highest concentration of naturally occurring opiates (endorphins) found in the brain. Since the nose is linked to the limbic system, it's possible that nose-stroking stimulates the production or distribution of these calming opiates.

The nose also has remarkable touch sensitivity. The tip is one of the most sensitive spots in the body. Animals seem to sense that touching nose tips is a good way to check each other out. It serves as a friendly greeting for members of the same species. Certain human cultures, such as the Eskimo, use nose-rubbing as both a greeting and kind of kiss.

We could certainly learn something from them. Nose-rubbing is a simple, gentle intimacy which creates a nice, warm glow inside. You can rub back and forth, or make little circles around each other's nose and face. One partner can nose-rub the other's lips, while the second responds by kissing and licking the first's nose. A wonderful sensation can be had from enclosing your nose completely in your partner's mouth, while he gently caresses it with his tongue. Many creative combinations come from alternating nose-rubbing with regular kissing and tongue-play.

An intriguing, but little known fact about the nose is that it contains the same erectile tissue found in the erogenous zones. This spongy tissue is located right beneath the mucus membrane. During sex play, the nasal erectile tissue becomes engorged with blood, causing the nostrils to expand. The nasal lining has a sympathetic reaction to stimulation of the erogenous zones. During long bouts of continued sexual contact, the nasal lining can become clogged up and chronically engorged, a condition doctors call "honeymoon nose" because it commonly occurs with newlyweds.

The nose may have other links to sex. No other animal has a protruding external nose like man's. Animals generally have simple openings which let the air pass through to the internal nasal cavity. In *The Naked Ape*, zoologist Desmond Morris speculates that our lengthy nasal appendage may have evolved as a phallic substitute. In the primitive past, our ancestors switched from rear entry "doggy-style sex" (prevalent among primates) to face-to-face positions. Morris reasons that the phallic-substitute nose provided a sexual inducement for the frontal posture. (Along the same lines, he speculates that fleshy female breasts may be buttock substitutes.) Whether or not that's correct, the sympathetic link between nose and penis/clitoris is a fact.

Another pleasing quality of the nose is that it merges two basic senses, smell and touch. When you nose-stroke sensitive skin like the face, you get the twin benefits of smooth touch stimulation plus a pleasant scent. There are few sensations more pleasurable than nose-rubbing and sniffing clean, healthy human skin. The best enjoyment comes from burying your nose in round, cozy places like the cheeks, nape of the neck, breasts, tummy, and thighs. The nose can be an effective touch-giver, although you may have to practice to get a soft, light touch. Your lover will enjoy tender nose-strokes and sniffs across the face, under the neck, down the shoulders, across the breasts, and over the back. You can actually nose-tickle your lover's tummy, sides, thighs and buttocks. A particular delicate nose might even be able to massage a woman's clitoris.

Some final nasal delights: light finger-tapping up and down the nose; the same using a feather or fragrant flower; putting your nose in the plush coat of a cat or rabbit. Cats, in particular, have a sweet, perfume-like scent on the back of their furry necks, a double-barrel smell/touch treat for your eager nose.

Nose rubbing   *Nose stroking*

Nose stroking

CHAPTER NINE: OTHER SENSITIVE SPOTS

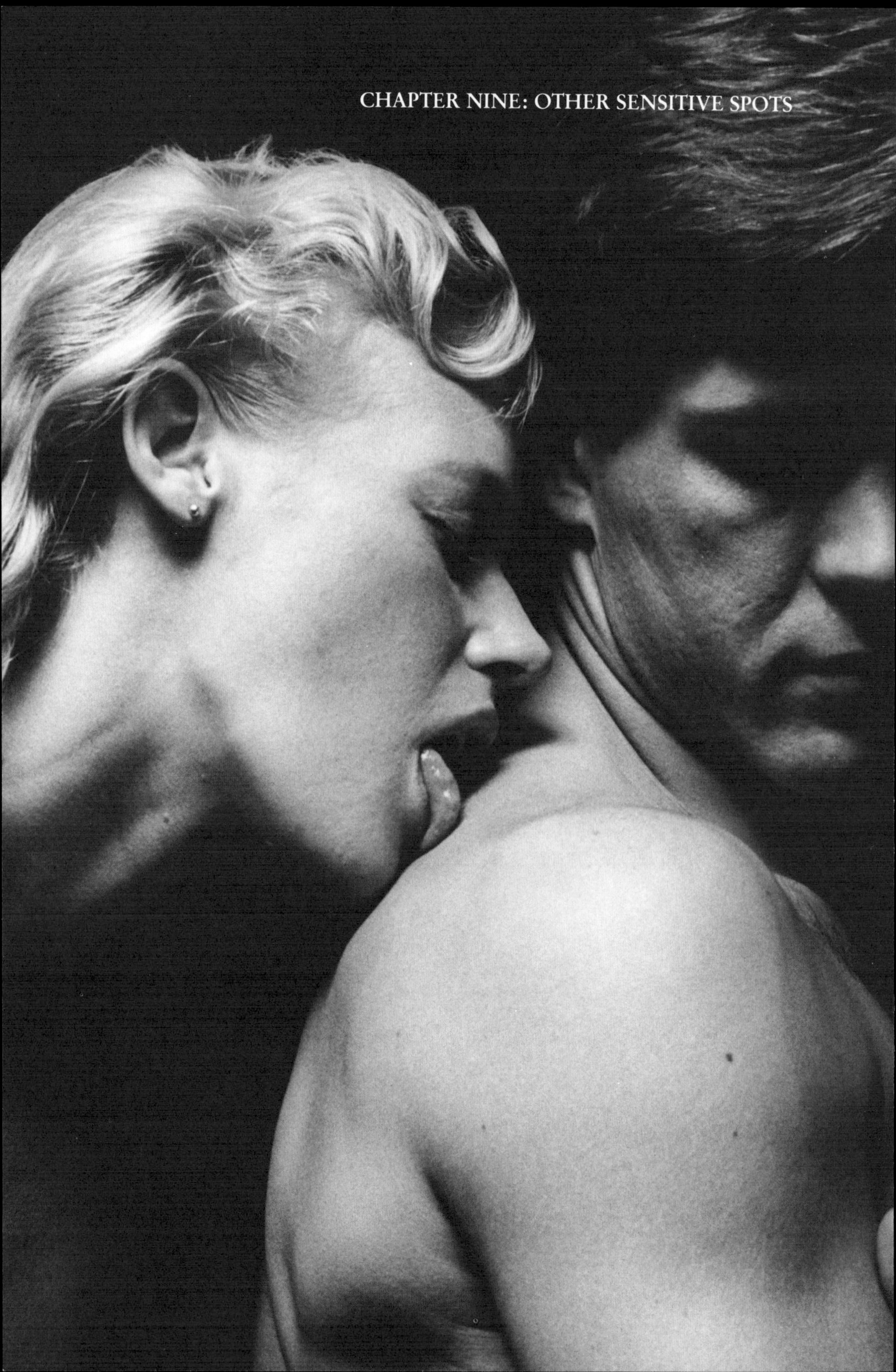

The Erogenous Zones These areas — which include the genitals, breasts, and mouth — have a dual sensitivity to both touch and sexual stimulation. This leaves room for lots of creative interplay. You can literally drive a lover up the wall by skillfully switching from light touching to direct erotic play. For instance, gently touching around the erectile tissue of the breasts (aureole) leads to a high pitch of anticipation for the fondling of the nipple itself. Keep your lover completely in the dark about when you intend to shift from innocent touching to sexual stimulation. Softly stroke the breast area and the surrounding chest in a casual, nonchalant way. Then suddenly caress the nipple between fingertip and thumb until it hardens. Replace your hands with your tongue to enhance the experience. Then just as suddenly, move back to innocent touch. Do this several times and check out the reaction.

Nipple play

The nipple play shouldn't be limited solely to women. We naturally associate breast stimulation with females, but males also have sensitive nipples. They'll grow hard from the right stimulation, and will trigger a sympathetic response in the penis. Switching from touch to erotic contact also works in the genital area. You might start with light strokes of the tummy, inner thigh and pubic area, then move into direct sex strokes on the penis/clitoris. Up the frenzy with some oral action on the sex organs, then calmly return to ordinary touch. Repeat the process if your lover can survive another round!

A variation of this switch game is moving from touching to tickling to erotica. A few areas, such as the nipples and buttocks, are highly sensitive to all three. You can combine the three movements any way that gives you the best response. However, keep in mind the likelihood that your lover will demand an orgasm break from this tactile torture. We can all handle just so much stimulation.

Speaking of stimulation, don't forget the mouth. This spot can be considered the great "touch corrupter" because it almost automatically turns innocent touch-play into sex action. Wherever they travel, the lips impart the kiss of sex. All body parts react to the soft, wet sensations of mouth play. You shouldn't limit the lips and tongue to your lover's mouth and genitals. For example, try slow, rhythmic tongue-strokes on your lover's shoulder. The mouth is both an excellent touch-giver as well as receiver. You'll both enjoy light, circular movements across the lips with the fingertips or tongue. The tongue itself can be massaged between the fingers and thumb, using a light, pressing action. We're all familiar with the delights of mouth-to-mouth play like lip-kissing and tongue-caressing, but that transcends touch into the realm of pure sexuality.

The Feet: The soles of the feet are among the most sensitive tactile spots on the body. Unfortunately, the feet are also one of the most neglected body parts. We abuse them constantly by wearing constricting shoes and walking on stiff, artificial surfaces like concrete. The feet bear the entire weight of our frame, so they certainly deserve better care. Our ancestors were aware of this need; witness the many references in *The Bible* about anointing the feet.

Our feet were built for barefoot strolls on natural surfaces. Modern living causes them to become soft and undeveloped, so barefoot walks on pebble-laden land can cause considerable discomfort. We still, however, enjoy the feeling of lush grass under our feet. To regain lost, sensitivity, try walking barefoot whenever possible, even if only indoors. This will also help eliminate foot odor, which is simply a perspiration problem caused by poor ventilation in shoes. You can also walk barefoot on soft, natural surfaces like sandy beaches and grassy fields. After awhile, you'll be able to move to harder terrain like rock or bare soil.

Like the hands, the feet are supposed to contain many pressure points which act as reflexes to different glands and organs. Walking on natural turf helps to stimulate these spots. Foot massage techniques (described in the massage and health sections later) offer another way to directly work on these pressure points. A foot massage can also soothe and relax the entire body. Fragrant oils add a nice touch. A similar relaxing effect comes from soaking sore, tired feet in a bath solution. They also respond well to the slapping and tapping techniques discussed in the section on "sensitivity strokes."

The soles of the feet form the prime focus for foot-touching. Soft, finger-tip strokes will send goose-bumps up and down the spine. You'll also trigger a tickle-response, since the sole is a top tickle spot. With their thick padding, the soles also react nicely to a strong pressure touch using the thumbs or knuckles. The tops and sides of the feet can be worked vigorously with the entire surface of the hands, employing a variety of pressures.

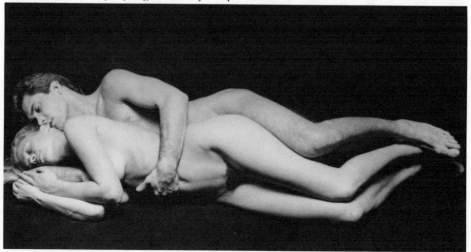

Foot-cuddling in spoon position

The toes are capable of giving as well as getting touch, but they're too clumsy to impart really sensitive strokes. Your feet can, however, give effective touches to your partner's feet. Pressing your feet together is a nice, soothing sensation for both of you. This can be done sitting on the floor facing each other or as part of the "spoon" position in cuddling, where one partner is curled around the other's back. In the spoon, the one in the back curls the top of his feet against the front partner's soles. The front partner then wraps his toes over the back one's toes, holding them nice and snug. This foot-cuddling gives both of you peaceful, relaxing sensations.

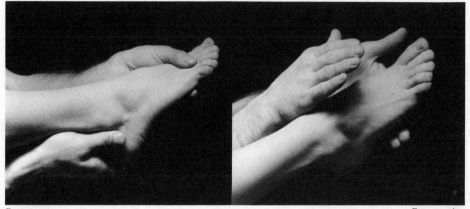

Foot massage   *Foot tapping*

The Trunk: Few body spots are oblivious to touch. However, areas do vary in tactile sensitivity. For example, well-muscled parts need deep pressure movements for maximum stimulation. On the trunk, these include the neck, shoulders, chest, abdomen, back and buttocks. Kneading, rubbing, percussion and other pressure movements are needed to work out trapped tensions in the large muscle group.

Feathery fingertip touches titillate the skin surfaces on the trunk. This can lead to both erotic and tickle stimulation. Especially receptive spots include the neck, breasts, tummy, sides, buttocks, underarms and pubic area.

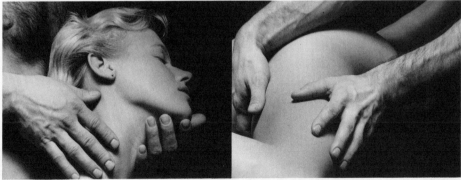

Neck rub   *Thigh rub*

The Limbs: The limbs also divide into deep and light pressure sensitivities. Muscular spots like the thighs, calves and upper arms require deep pressure to soothe embedded muscle stress. Virtually the entire skin surface of the limbs responds well to feathery touches, some of the best spots being the thighs, upper arms and calves. Good tickle targets include the knees, inner thighs and upper arms.

As touch-givers, the limbs' basic role is for wrapping, hugging and embracing your partner.

PART III: TOUCH TECHNIQUES: STIMULATION AND RELAXATION

The late 1960s saw an explosive expansion in human consciousness. People seemed to be pushing against limits in all areas—the political arena, sexual awareness, race relations, drugs, etc. Part of this process was a revolt against cold, abstract intellectualism—the calculating, rigidly logical rationalism which pervaded such important institutions as business, government, the military and education. The seekers of the Sixties were searching for ways to return to feelings, intuitions and bodily sensations, without the castrating tyranny of the mind.

And so we saw the birth of the "human potential movement" in its various forms: T-groups, sensitivity training, encounter groups and numerous others. Most of these efforts tried to break down the artificial barriers between people, including the touch taboos. Encounter group leaders understood that loosening touch restrictions paved the way for spontaneous communication and rapport. So they employed touch exercises as warm-up tools for verbal exchanges. In fact, the public came to define encounter groups by their touch tactics. The popular phrase "touchy–feely" was coined to identify them.

One of the prime movers in both the early and present stages of the human potential movement is the Esalen Institute in Big Sur, California. People working at Esalen pioneered exercises for awakening the senses, enhancing interpersonal awareness and increasing bodily attunement. Many of these methods are touch techniques which belong in this book. However, when first researched they seemed a bit puzzling—quite different from such familiar techniques as massage, cuddling and healing touch. The Esalen-type touching seemed simpler, less intense, almost playful. They were like preliminary forms of tactile activity, the perfect appetizer to prepare you for a full course.

The Esalen-type exercises are charming, childlike ways to whet the appetite for more intense touch. Since they spark skin sensitivities, they may fairly be considered "sensitivity strokes." While some of them are fairly intimate, most are innocent enough to practice with friends and even acquaintances. They are particularly useful to "break in" someone whose tactile sensibilities are undeveloped, and can also serve as "foreplay" for more involved touch experiences like massage and erotica.

Most of the credit for these sensitivity strokes goes to Esalen's Bernard Gunther and his excellent book, *Sense Relaxation*. Credit is also due Dr. Sidney B. Simon, an educational consultant who has used some of these techniques in classes at the University of Massachusetts.

Innocent Strokes

Back Rub: One of the simplest sensitivity strokes is the ordinary backrub. It can also be a real wonder-worker, because this area is a tension-pocket. Offer this anytime to a friend who complains of nervousness, stress, or anxiety.

The best position for a backrub is to seat your friend on a low chair or stool with his head and arms bent over a taller table. You will be standing behind him. Put several pillows under his arms for comfort. Gently, but firmly, knead and rub all the tense tissues on the neck, shoulders and back.

Arm Rub: Hold your friend's hand between both of yours. Press up the arm all the way to the shoulder, then back again several times. Vary the contact between light and moderate pressure. After finishing, don't do the other arm right away; first call your partner's attention to how deprived the untouched arm feels. As Dr. Simon says, "The contrast between the two in feeling is so dramatic that it rams home the point of how the untouched people in our society must feel most of the time — dead, uncared for, bland." (*Reader's Digest*, May, 1980).

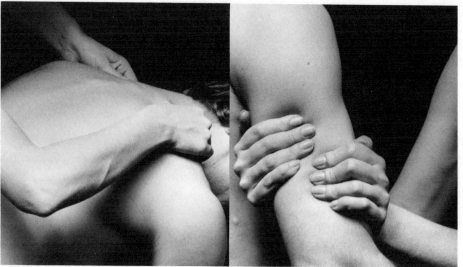

Back rub   *Arm rub*

Temple Touch: Seat your partner in a chair, while you stand behind him. Lean his head against your body and put your hands on his temples, one on each side. Give several minutes of light massage with the flat part of your fingers. This can relieve headache and tension.

Hand Hugging: Sandwich one of your partner's hands between the palms of your hands. Fold your fingers over for a good grip. Use moderate pressure. The idea is to warm up cold hands. If the hand is still cold, rub it between your hands. Then do the other hand.

Footwork: A good treatment for cold feet. Take your friend's feet in your lap and warm them up with your hands. Rub, knead and shake them until the blood flow heats them up. You'll find that eliminating cold feet helps your friend to "open up" and communicate better.

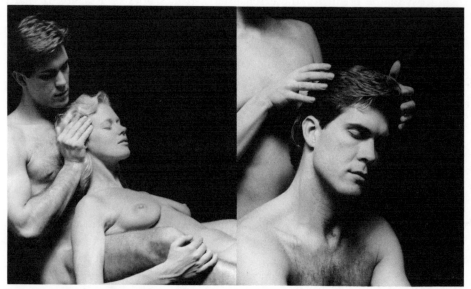

Temple touch   *Getting a head*

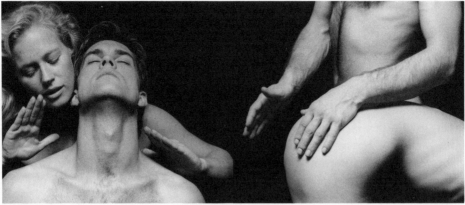

Shoulder slaps   *Back battery*

Getting a Head: Stand behind your partner, who sits with his eyes closed. Bend your
fingers and begin gently tapping all over his head, including the forehead. Tap with
both hands at the same time in a leisurely, even rhythm. Do this for a minute or so,
then lightly touch all the areas you've just tapped. Use the full flat of your hands to
gradually cover the entire head. Now move to the front of the face, and gently
explore it with soft caresses. You can vary this by running a flower across the face.

Shoulder Slaps: Have your partner stand in front of you, eyes closed. Flatten your hands
and begin gentle body-slaps on the top of his shoulders. Work over the shoulders,
down to the fingertips, and back again several times. Follow this with light
full-hand touches to all the spots you've tapped.

Back Battery: This time, do the slapping from the shoulders down to the backside. Your
partner can stand straight up, bend over, or lie down for this. It's all right to be
vigorous on the back, especially at the butt. Light spanking is often a good
stimulator! Work up and down the back a couple of times, then finish by gently
resting your hands on all the slapped parts.

A Leg Up: Have your friend lie down flat on his belly. Slap his legs from thigh to toes.
Don't miss the sides of the leg, nor the foot itself. Slap up and down a few times,
then complete the leg work with soft touches.

All Shook Up: Your partner stands in front of you, eyes shut. Put your hands on his and
begin shaking them simultaneously. Shake your way up the arms and shoulders,
down the front of the body and legs, all the way to the ankles. Do this for a while,
then let the person stand alone for several moments to enjoy the effects.

All shook up

Intimate Strokes The "innocent strokes" just described are preliminaries for the intimate strokes, which may be more appropriate for someone close to you. They are more intense, and can even be sexually stimulating. Some of these moves can be combined nicely with such sensual activities as massage and cuddling. Others are almost forms of sexual foreplay.

Backing Up: You and your partner stand with backs up against each other, eyes closed. The idea is to communicate non-verbally with your backs. Stand still, move, rub each other, respond to each other's movements. Then turn around and look into each other's eyes for several minutes.

Head Play: Seat your partner in front of you with his eyes shut. Softly touch his entire head and neck. Rub his scalp tenderly, stroke his forehead, caress his ears. Then move to his face and explore it in the same way.

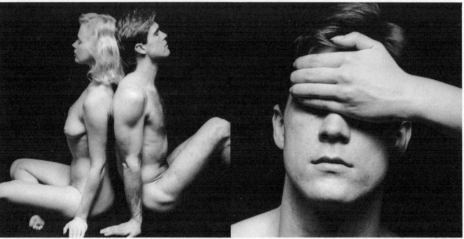

Backing up   *head play*

Being Upfront: Sit facing each other on the floor. Reach your hands out so your palms touch those of your partner. Close your eyes and maintain the silent palm contact for a while. Then, eyes still closed, use your right hand to explore your partner's left hand, while he does the same to your left hand. Feel the texture of the fingers, palm, back of the hand. Continue to explore up the wrist, the elbow, upper arm, shoulder. When you reach the top of the left shoulder, leave your hand there. When your partner finishes exploring your left arm, rest your left hand on his right shoulder. With hands on each other's shoulders, draw your heads together and just lean against each other for awhile. Now reverse the movements through to the beginning until only your palms touch.

Then stand up and embrace each other. Make sure your entire bodies snuggle together tightly. Eyes closed, just lean into each other for as long as it feels good.

Foot Frolics: Have your partner lie on his back, shoes and socks off. Sit down next to him, and start slapping his feet between both your hands. Include both the sides of the foot and the top and bottom. Finish with soft caresses of the entire foot. Duplicate the movements on the other foot.

After your partner has done *your* feet, both of you lie down on your backs, eyes closed, opposite each other. Bring your feet together, soles touching. Communicate non-verbally with your feet through rubbing, poking, wiggling and being still.

Being upfront (1)

Being upfront (2)   *Foot frolics*

Water Sports: A nice follow-up to footwork is a foot bath. Use a container filled with comfortably warm water, and tenderly lather and rinse your partner's feet. Do this slowly and thoroughly. Then dry them and annoint them with oil or lotion.

Another nice water sport is the head shampoo. Use a pitcher or flexible plastic spray hose as a water source. Wet down his hair, then rub in his favorite shampoo. Be exquisitely sensitive as you work on his scalp. Take plenty of time doing this. Rinse the suds off, then dry him with a towel. Comb his hair out.

The next step up in water sports is the full-body bath or shower. Take turns washing each other's bodies, head to toe. Take the time to do a thorough, sensitive job. Rinse and towel each other dry. This body bath can serve as a fitting prelude to the full-body massage discussed below.

Group Strokes

Pass-On: Six to eight people form a closed, tight circle around one individual, who should close his eyes and prepare to surrender completely to the group movements. He stands erect with feet together, then falls backwards. The person in the circle behind him will catch him, then pass him along to someone else. Pass him in different directions and at different speeds. The group then encloses the individual in the center of the circle, each member pressing against him. After a minute, the group slowly withdraws, leaving him standing alone. Each person in the circle should get a turn at being passed.

A Real Lift: One person lies on his back, eyes shut, in the middle of the group. All members put their hands under the individual, and they slowly lift him off the floor. He should be lifted to shoulder height, and maybe even overhead. The group can rock him, swing him, run around with him, or toss him in the air and catch him. Then he's slowly lowered to the floor, and the group covers him with their hands for a few moments. Let him lie there quietly for a while before moving on to the next member to be lifted.

Pile Up: Each member closes his eyes and crawls towards the center of the room. In the center, a pile of crawlers will form. Everyone just keeps adding to the heap. When there's no more movement, everyone stays still for a few moments, savoring the feeling of the people they're in contact with. Then slowly unpile.

Gunther Sandwich: The group members all lie down next to each other. Then they all turn and roll over on their left sides. Each person puts his arm around the person in front and they snuggle up close, lying together for awhile and enjoying the collective warmth. Then the last person in the line gets up and moves to the front snuggle-position. Continue this until the group returns to its original position, pausing between each new formation. Then slowly unwind.

Many of the other movements described above for *two* people can also be done in groups. For example, two or more people can "slap" a third person. It's just a question of dividing up the body to best fit the number of available hands. Similarly, a single person can perform most of the tapping, slapping and touching strokes on his own body. The head, legs and feet can easily be touched with both hands, while the shoulders and back can be reached by crossing your arms at the chest. These exercises are designed to awaken the senses, so they work well as a morning fresher-upper for someone who's alone.

Gunther sandwich for two

CHAPTER ELEVEN: MASSAGE (PART 1)

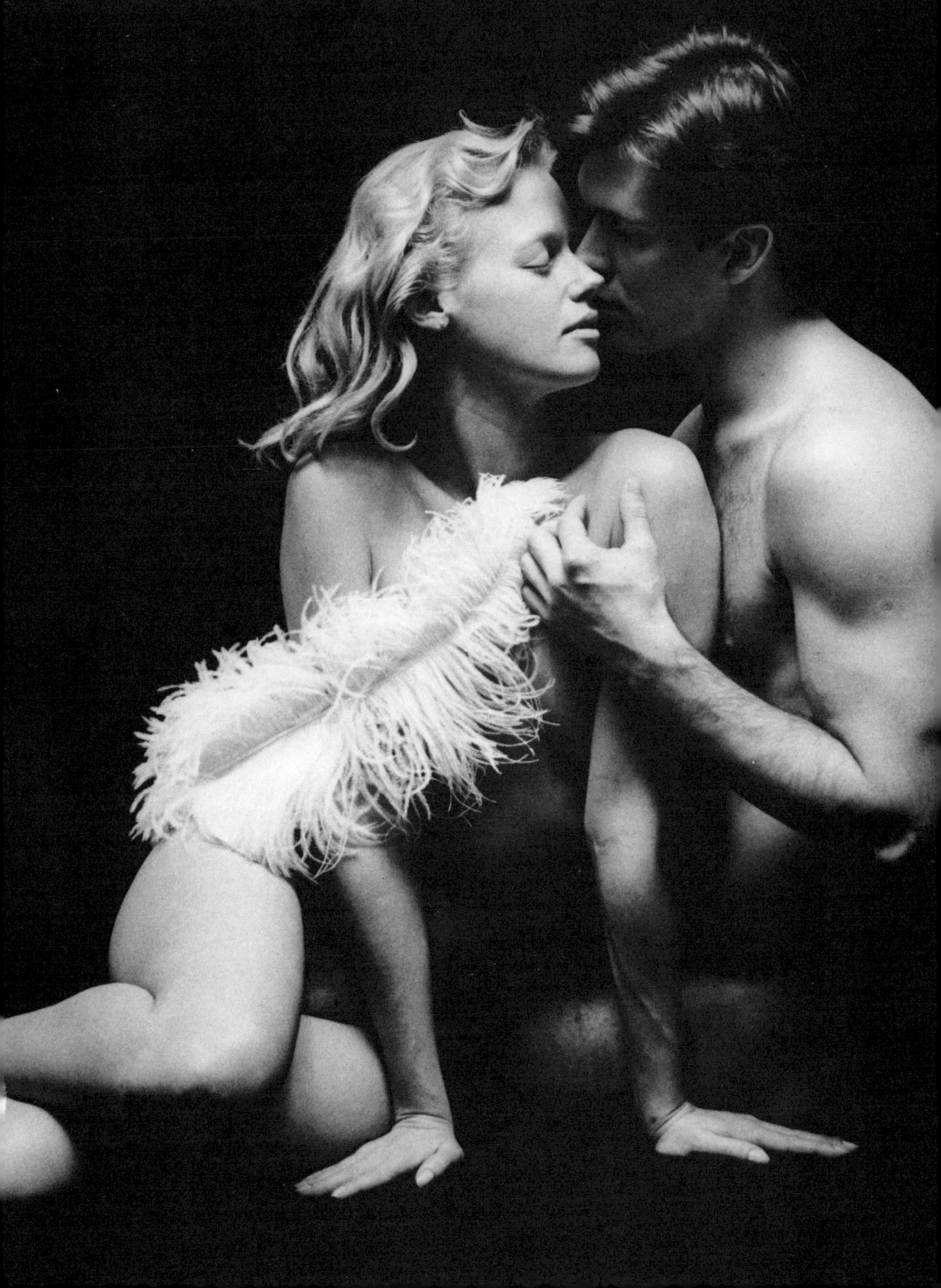

Introduction A good massage is one of the most pleasurable experiences anyone can have. Massage can be a total touch orgy — every inch of the skin gets pressed and soothed by nurturing hands. The internal organs, joints, even the bones receive their share of stimulation.

It's hard to overstate the joys of massage, because the benefits are so numerous. The lift is psychological as well as physical. What better way to express affection, caring and love than by skillfully touching your partner all over? Unlike most sex, there's a wonderful selflessness about massage — one partner totally gives, while the other totally surrenders. In sex, mutual pleasure — simultaneous giving and receiving — is the norm, and this can sometimes interfere with the full gratification of one or both partners.

Mixing the giving and getting roles can mean that neither partner gets total attention. Many lovers feel guilty about just laying back and passively surrendering to pleasure. This mutual pleasure-obsession leads to excessive ego-involvement in the bedroom. The quest for "simultaneous orgasm" puts a bigger premium on performance than on pleasure. Not reaching this overrated goal leads to feelings of doubt, frustration and failure, which are hardly conducive to further sharing in other aspects of the relationship.

Modern sex therapists often recommend that their clients practice "solos," in which both partners take turns being completely catered to sexually by the other. The massage experience is similar to sexual solos. One partner totally concentrates on administering pleasure, while the other fully succumbs to it. Each one gets to completely express his passive, accepting part as well as his active, giving side. This creates the ideal conditions for trust, openness and sharing.

Similarly, the massage-giver learns to appreciate the joy and delight his hands can bring. He gets a powerful sense of accomplishment from being the instrument of such tactile rewards. Meanwhile, the massage-receiver luxuriates in total, trusting surrender. Being completely catered to expands his depth of pleasure-experience, making him in turn a more effective pleasure-giver.

As Gordon Inkeles has said in his book, *New Massage*, pleasure itself can be therapy, particularly in our anti-tactile culture. Unfortunately, massage has historically been a victim of these prejudices. Just as touching became linked with sex, so massage became connected with prostitution. The "massage parlor" stereotype is still strong with much of the public, but it is slowly losing its grip. In fact, there are signs that we are in the first stages of a massage renaissance. Books and articles on the subject have multiplied in recent years. The Swedish Institute in New York, the oldest school of massage therapy in the U.S., reports that its enrollment has *tripled* since 1975.

This renewed interest should help return massage to the respected role it once held. Massage is one of the oldest forms of therapy known to man, possibly beginning in China over 5,000 years ago. The ancient Greeks and Romans used massage as part of their physical education programs. In 400 B.C., Hippocrates, known as the "Father of all Healing," employed massage to treat his patients.

Actually, it's only fairly recently that the medical profession abandoned massage as a healing tool. There are early 20th Century medical texts which prescribe massage movements for certain conditions. However, massage no longer fits in with the detached, professional demeanor of modern doctors. It's also out of place with our growing dependence upon drugs to treat all discomforts.

Despite conventional medical attitudes, massage remains an effective drugless therapy. There are few body parts which don't benefit from it. Circulation is enhanced, skin and muscle tone are improved. Digestion is speeded along, fat deposits are loosened. Fatigue-producing chemicals are cleared away, a benefit of exceptional interest to athletes. In fact, organizers of the New York City Marathon have invited massage therapists to join doctors and physical therapists in administering to runners at the finish line. They recognize how dramatically massage serves to accelerate recovery from exhausting exercise.

Massage enhances the flow of lymphatic fluids, so that the tissues become better nourished. Scalp massage stimulates the sweat glands, thus moistening and toning the hair. Massage has a profound effect on all stress-connected conditions, too. The sensitive nerve endings of the skin are soothed and relaxed. Specific tension-pockets like the neck and shoulders get substantial relief. Tight, stressed muscles cause pain, and massage helps to loosen them. Regular massage sparks the central nervous system, leading to livelier overall functioning. In many respects, the effects resemble those of vigorous exercise, but *without* excessive work for the heart. This is particularly useful for disabled people, since massage slows down atrophy in body areas that can't easily be moved.

It also alleviates certain ailments and injuries. Headaches, stomach "knots," cramps of the limbs, aching feet, and sore, overworked muscles all respond to different massage movements. Congested lungs, insomnia, stiff joints and muscle sprains can also be relieved. Certain psychological problems can be candidates for massage work. In *New Massage*, Inkeles correctly observes that massage can serve as psychotherapy. People with sex problems often cut themselves off from all pleasure, and massage can reverse this. Actually, many emotional disorders stem from poor body self-image, especially in our non-contact culture. Massage could help people with conditions like schizophrenia connect with their bodies, feelings, and internal sensations. How different would our mental hospitals be if massage therapy replaced such violent treatments as electroshock, lobotomy and mind-sapping drugs?

Massage obviously has many applications. Apart from all the specifics, however, the real bonus of massage is a general feeling of well-being. A full-body massage soothes, stimulates and enlivens you all over. Put simply, you feel great! This wonderful sensation makes massage very versatile. It can be used as stimulating sexual foreplay, as a relaxation aid, or as therapy. It can be combined with other therapies, or with exercise or meditation. The creative combinations are countless.

As massage moves further from its "dark age" into new respectability, other intriguing applications will doubtless be found. It's interesting to speculate what American life would be like if massage spread to all homes. Imagine all parents massaging their kids and all lovers massaging each other! We might have less delinquency, violence, frustration and anger. And in its place more calmness, peace and love. The world definitely owes massage the chance to work its soothing magic.

There are many schools of massage, but two basic styles are practiced today. *Western*, or Swedish, massage is familiar to most. It's the kneading, pressing, and rubbing style commonly seen in health clubs, spas, etc. The Swedish style, which is what most people think of as massage, will be described in this chapter and the next.

The second basic style is *Oriental* massage, also known as acupressure and Shiatsu. This is based upon the same principles as acupuncture: Energy is said to flow through the body along invisible pathways called meridians. Stimulating certain points on these paths will alleviate different ailments. A Western system called reflexology also uses pressure points in a similar way. Since Shiatsu and reflexology massage are, first and foremost, *healing techniques* (rather than sensual experiences), they will be discussed in Part IV of this book.

Easy Massage A truly professional-quality massage is complex. There are many movements, variations, and accessories needed for a topnotch job. This kind of the quality full-body massage is a priceless, irreplaceable touch technique. But it *is* relatively elaborate, equivalent to a multi-course tactile dinner, candlelights, the best china, and all.

We also need a massage that's right for a light touch-snack. Something fairly effortless, not too time-consuming, but which still works the body all over. Something which doesn't require lots of special rules, technical moves, nor lengthy preparations. Actually, we all occasionally experience a variation of this "easy" massage—the ordinary back rub. What the easy massage does, in a sense, is extend the back rub over the entire body.

Easy massage is especially useful at these times: before lovemaking; when someone is very tense or anxious; before getting up in the morning, or when going to sleep at night. In all these cases, it serves to soothe skin and nerves, relax tight muscles, and make the mind mellow—benefits which are useful whether you're about to go to sleep, tackle a new day, make love, run a marathon, or attend a stressful job interview.

Since easy massage strokes are mainly light and gentle, the active partner—the one *doing* the massaging—shouldn't end up being exhausted. This means you can do the manipulations wherever you happen to be lying—in bed, on a plush carpet, etc. There's no need for oil nor alcohol. Ideally, you should both be naked. There are only a few things to remember: make sure your partner is warm; keep all noise, distractions, conversation and interruptions to a minimum; don't work on injured or diseased areas. Some soft background music creates an appropriate mood.

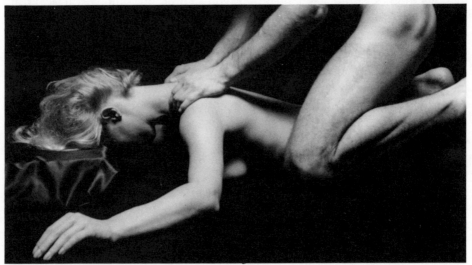

Neck rub

You can start anywhere on the body, although many prefer the back of the neck. Tensions commonly accumulate there, so it's a good place from which to launch the relaxation process. Have your partner lie on his stomach. Place both of your hands on the base of his skull, thumbs touching in the center and fingers extending around the sides of the neck. For the neck rub, it's probably best for your partner's head to face straight down, chin against a table, pad or other surface so that the neck muscles are prominent. For all other movements, the head should be comfortably rested on either side. The active partner should take the most comfortable position near his friend's body — kneeling to the side, squatting, gently straddling — whichever works best.

Move your hands up and down the neck, stroking, rubbing and kneading all the knotty muscles. Use the flat part of your fingers, palm and thumbs, whatever best fits the contours of the area you're massaging. Search the muscles for pockets of tension and work these out with deeper pressure. Go lighter on thinly-muscled spots and around bones. Easy massage relies mostly on your intuitions for the right pressure and movements. As long as you're basically gentle, you shouldn't have any problems. Look to your partner for cues — "ouches" and flinches say you're too rough; sighs of satisfaction tell you that you're on target, and you may wish to linger on such spots for several extra moments.

There are no set time limits for any movement. Just do what feels right to you and your partner, generally spending a few minutes on each spot. Your only eventual goal is to cover your partner's entire body with pleasing touches. While you're concentrating on the neck, make sure you work the muscles on the base of the skull, around the ears and on the sides of the neck. After several manipulations up and down the neck area, use your fingertips for an invigorating scalp massage all over the head. Move in nice, smooth circles, taking care to avoid jerky, uneven motions.

Then travel down the neck again to the shoulders. Your left hand rests on the left side of the neck and your right hand on the opposite side. Work outward with each hand towards the end of each shoulder. Move towards the neck and away again, giving all those knotty, tension-filled muscles plenty of good kneading. Cup your hands around the outer part of the shoulders and do some deep, circular kneading.

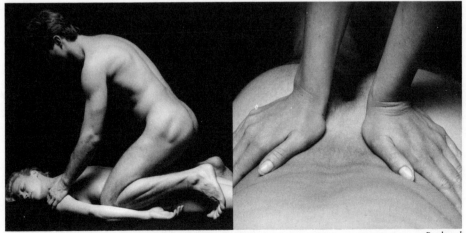

Shoulder rub   *Back rub*

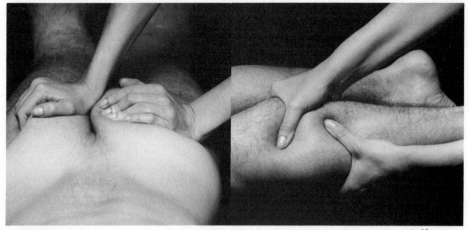

Buttocks kneading   *Calf massage*

After several minutes of shoulder work, travel down the back. Use the full flats of your hands to press, knead, and stroke up and down the back, and around the sides. Be light around the spine, but otherwise deep pressure is okay. There's no such thing as spending too much time on the back. Work all tension-pockets thoroughly, use your intuitions, and heed your partner's nonverbal feedback.

Next spot is the buttocks. Give the cheeks a vigorous workout, with deep kneading and pressing. Then travel to the back of one thigh, where you'll use both hands. Press up and down with your palms, and knead with your thumbs. Repeat these moves down the back of the knee and calf. Use thumb and forefinger to manipulate the slender area of the ankle. When you reach the foot, cradle it on one of your legs or on a pillow. Grab the foot between both hands with your thumbs on the center of the sole. Press deeply up and down and around the sole with your thumb and fingers. Repeat the whole process on the other leg, starting again at the back of the thigh.

Now have your partner turn over, and continue with the front of the foot. Stroke all the toes with your fingers, then use your full hand to press the front and sides of the foot. Work up the ankles and shins with your full hand, palm and thumb. Use circular, rhythmic moves on the knee. The front and sides of the thigh get massaged with thumb-kneading, palm-pressing and full-hand stroking. Then do the same for the other leg, starting with the foot again.

When you complete the other leg, work lightly over the lower abdomen and belly. Stick to gentle strokes here, in order not to upset digestion. Then move up and around the chest area with both hands. Use circular, rhythmic motions. Use fingertips to rub sensitive spots like the breasts and solar plexus. Massage the front of the shoulders with the flat part of your fingers, switch again to the fingertips for the area around the clavicle bone (top of the chest). Move from the chest to one of the arms.

Position yourself to the side of your partner so that you can rest his arm across your leg. Encircle the top part of the arm with both hands. Knead and press with your fingers, thumbs and hand (using medium pressure) up and down the arm, making sure to include all the muscles, front and back. Knead the elbow with your whole hand, then use thumb pressure to work down the forearm, while the rest of your hand presses the back of the lower arm. Thumb and finger pressure are best for the wrist. Your partner's hand should be massaged all over, with deep

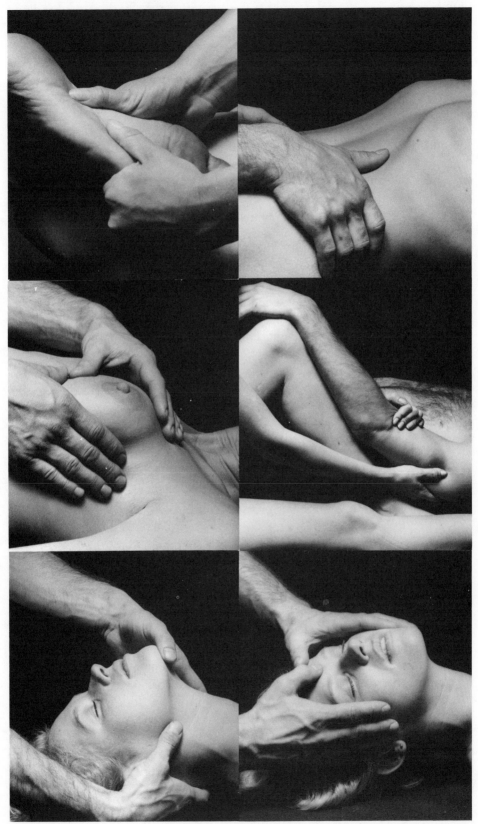

Foot work; Belly rub; Breast massage; Arm kneading; Jaw massage; Facial strokes.

thumb pressure on the palm, and gentler strength on the back of the hand and fingers. Each finger should be kneaded between your thumb and fingers. Repeat the entire process for his other arm.

After finishing both arms, gently rub the muscles on each side of the windpipe. Also include the muscles on the sides of the neck, using light fingertip pressure. Move up to the jaw, massaging it with gentle upward pressure from your thumbs. The chin and the muscles around the mouth, eyes, ears and forehead all respond to easy presses of the fingertips and base of the thumb. Just follow the contours of the face with your fingers. Curl around the mouth, go up and down the cheeks, across the forehead, around the eyes and temples, over and behind the ears. Rest two fingertips *softly* upon the closed eyelids. Place the thumb and forefinger on opposite sides of the bridge of the nose, and give light pinches down to the tip. Use a similar pinching technique up and down the ear.

Finish the face with finger strokes starting at the forehead. Place the flat part of your fingers on your partner's forehead, hands facing each other. Then softly stroke down the forehead, over the temples, across the face, over the ears, down the jaw. For a final touch, give another brisk fingertip scalp-rub.

That's it. Let your partner just lie still for awhile, luxuriating in tactile delight. If he falls asleep, let him be — that's a sign of a deep-seated need for rest. Don't be over-anxious for your turn; it will come soon enough. Your partner should gratefully return the favor.

There's nothing especially scientific about the easy massage. Some might even object to calling it a massage. Maybe it should more properly be called a "full-body rub" or a "total touch experience." Whatever the label, the easy massage is a great way to escape from the "touch closet" so many of us get trapped in. It's rare that we ever receive such sustained bodily attention.

The really great thing is that it's so simple. It can be done anytime, almost anywhere, and with no special expertise or equipment. It probably won't provide all the therapeutic benefits of a professional-quality massage. Nonetheless, it *will* give plenty of the hearty well-being provided by all rich touch experiences — a real "contact high."

Participatory Massage Another style of massage is similar in some respects to easy massage. It also requires no special expertise nor technical knowledge. Like easy massage, it relies mainly on intuition, and it doesn't wear out the massager. That's where the similarities end. Jonathon Daemion, a holistic psychotherapist, developed this new variation. He calls it "co-reflexive touch therapy," but "participatory massage" is a simpler way to say the same thing.

Daemion's method allows greater involvement for the person receiving the massage. The massager begins with a circular motion somewhere on the receiver's body. He then invites the receiver to move his body according to the rhythm of the massage movements. The receiver does this, shifting his body with the rhythm. The movements change, as each one adjusts to the other. "It becomes a dance," Daemion says, "a natural movement which doesn't drain the massager." Another advantage is that the receiver can assist the massager in hitting particular pockets of tension he's working near. The receiver can shift the sore spot in line with the massager's hands.

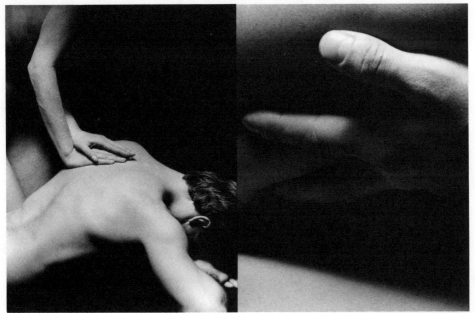

Participatory massage   *Love massage, thighs*

At a demonstration of Daemion's technique, he randomly picked someone from the group and had the fellow lie on his stomach. He began a circular stroking motion on the fellow's back. The man soon arched his back to match Daemion's movements. Daemion then shifted his own motion slightly to better meet the man's actions. In a short time, Daemion's hands were moving nicely to the same beat as the gyrations of the receiver's back. It was indeed a massage-dance.

Love Massage Massage and love-making are a marvelous combination. Good sex generally depends upon a mellow state of mind, and that's exactly the magic massage works. The soothing, hypnotic haze of a solid massage is a fitting prelude to the titillation of sex play.

A love massage is begun after the regular massage has been completed. Basically, it consists of a series of soft strokes on very sensitive body spots. This stimulates your partner in the same way as foreplay. Start with light fingertip strokes across the entire body, head to toe. Then shift your attention to increasingly sensitive parts. Make little circles across the palms of your partner's hands, soles of the feet, and stomach. Trace your fingertips across the edge of the ear and lips. Continue your feathery touches along the buttocks, tailbone and inner thighs. Run your fingers between the toes. For an exciting variation, blow tiny bursts of air all over the body.

The breasts deserve special attention. Cup your hands over both of them for a while. Then move your fingertips around the fleshy part. Avoid the nipples at that point. Eventually, move right over the nipple with one finger. Move back and forth or circle around the nipple, barely touching it. Repeat this breast-nipple combination several times, then—without warning—rub the nipple between thumb and forefinger until it becomes hard. Do this for a while, then switch back to light strokes. Ignore any protests, because part of the fun is tantalizing your lover. You want to build to a craving for hard erotic action. By the way, this nipple-play shouldn't be restricted to females; it also works nicely with males.

The anticipation game also applies to the genital area. Gently rub and stroke around the pubic area and inner thighs, then narrow your circle. With a single finger, trace along the line between the anus and genitals. If you're working on a woman, the next move is to stroke your fingertips along the outer lips of the vagina. Moisten a forefinger with saliva, and softly move over the clitoris area. Pull the skin back and barely touch the clitoris itself. Be exquisitely gentle here.

After some time, shift back to more innocent strokes. Play with the pubic hair for a while, then gradually work back to the vaginal lips. From there, put your forefinger inside the vagina, massaging the inner walls. Then return once again to less erotic strokes.

If your partner is a man, start the genital play with light fingertip movements around the testicles. Do this for awhile, then start tracing up and down the phallis with your fingertips. Now cup the penis in both hands, and give a series of gentle mini-squeezes up and down the shaft. This should help pump the organ up to good size. The final touch is to trace tiny fingertip circles on the erect head. From there, switch back to less intimate touching.

If you want to boost the love massage frenzy to the nth-degree, add little bites and tongue-licks to all the movements above. A final tantalizing touch is moving a silk scarf, furpiece, or large feather slowly over the genital area.

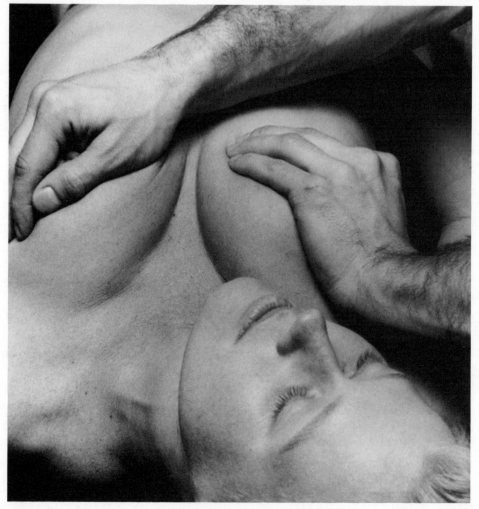

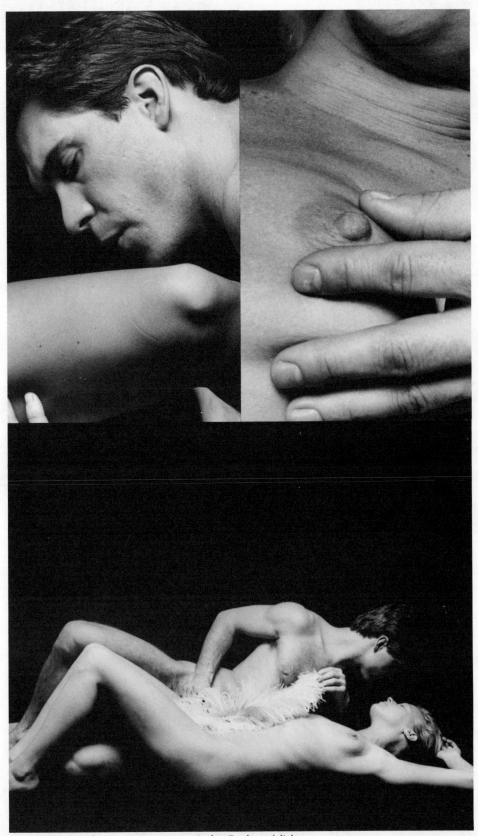

Love massage, air bursts; Love massage, nipples; Feathery delight.

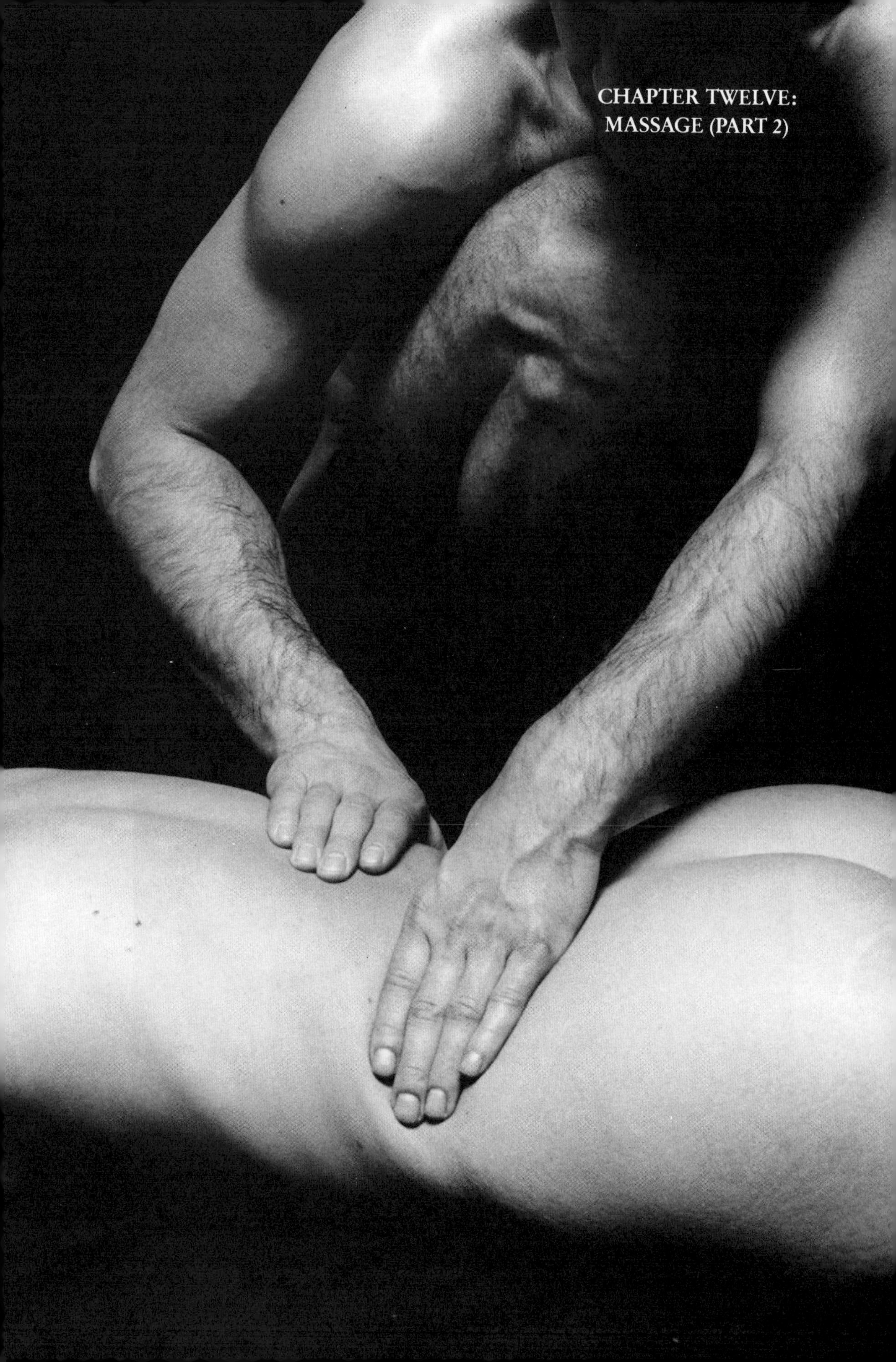
CHAPTER TWELVE:
MASSAGE (PART 2)

Full-Body Massage The full-body massage is a true sensual feast. It takes a decent amount of energy and preparation, so it's a very special tactile gift for your touch-partner. But it is well worth the effort. Your partner will likely appreciate the love and devotion, and amply pay you back in kind.

It's hard to overestimate the benefits of a full-body massage. Such physical rewards as relaxation and stimulation are obvious, but the psychological payoff is just as important. Having someone completely cater to your bodily pleasure is a rare, but powerful, treat. It's difficult to imagine other situations where you will receive such total, absorbing care. You'll end up with a "blissed-out" consciousness. When you understand the feeling, you'll be motivated to share it with others. It's gratifying to know your hands can produce so much potent pleasure for other people.

If you have never had a professional-quality massage, you owe it to yourself to look up a local massage therapist. The usual fee of twenty to thirty dollars an hour is a wise investment in yourself, especially when you consider how much money is tossed away on trivial, fleeting entertainment. A massage a month should be the minimum for anyone working in a modern, high-pressure environment.

Experiencing a professional massage will give you enhanced insight into the techniques presented in this chapter. But another good way to learn is to work closely with your touch-partner. Practice the moves a few at a time on each other. When it comes to massage, two heads are clearly better than one. Share your impressions. *Communicate*. Give your partner feedback on how he handles the various pressures, movements and methods.

Don't tackle too much at once. Thoroughly learn the techniques for one body spot before moving on to another. And remember, have fun while learning. Your goal remains the same, before *and* after you master the movements—giving and receiving pleasure.

If you initially decide to learn the techniques by yourself, don't forget that the key to effective massage is the sensitivity of your hands. They should be mobile and dextrous, traits you can develop by practicing with musical instruments, card tricks, or juggling. You can also work on specific massage movements with a well-stuffed pillow or cushion. But the next best thing to practicing on someone else is to use your own body. With the exception of some tough-to-hit spots on your back, you can reach most areas with your own hands.

Preparations There are a number of things to keep in mind before starting a full-body massage. First, you should know when *not* to massage. *Never* massage any area with these conditions—open sores, fractures, sensitive veins, bruises, tumors, joint inflammations, cuts, serious bruises. Nor should you massage anyone suffering from a fever, acute stomach problem or skin disorder.

Here are some additional tips:
* The person being massaged should always be kept warm and comfortable. Room temperature should be about 75 degrees.
* Any body part being massaged should be well-supported.
* Focus fully on the massage and on the person being massaged.
* Keep distractions and interruptions (like a ringing phone) to an absolute minimum.

* Before starting, ask your partner if there are any particular tension spots you should pay special attention to.

* The receiver may be a bit uptight in the beginning, but will relax as you proceed with the movements.

* Maintain a consistent, even rhythm in your movements. Avoid jerky, random motions.

* Try to constantly keep contact throughout the massage. It's unsettling for the receiver to go along stretches without feeling your hands.

* Don't give a full-body massage on a bed. It's too exhausting for the massager.

* Quiet is the rule during massage. Long discourses about what you're doing interfere with full sensual appreciation.

* Find out how you're doing through non-verbal cues like sighs, groans and muscle movements. As you practice, your hands will become increasingly sensitive to the receiver's feelings.

* Don't massage anyone after he's just eaten. Give him time to digest the food.

* Make sure the massage room is free from harsh, direct lighting. Use softer lighting like low-wattage colored bulbs, candles or oil lamps.

* Always remember to balance your movements. If you massage one side of the body, do the opposite as well.

* When receiving a massage, empty your mind of all thoughts. Let your body hang loose, limp and relaxed.

* Both receiver and massager should get into the right frame of mind by practicing deep breathing before beginning.

* The receiver should remove all jewelry, glasses, contact lenses, and as much clothing as he's comfortable with. It doesn't hurt for the massager to be nude as well; at a minimum, his arms should be free of restrictive garments.

Since the skin is going to be rubbed all over, it's wise to clean it before starting. A warm bath, shower or sauna should do the trick. In *New Massage*, Inkeles suggest making this cleaning a part of the massage experience. With a shampoo brush, loofah or bath mitt, the massager should soap the receiver's body all over. Use smooth, circular strokes. A flexible shower spray should be used for rinsing. The best place to bathe someone is on a bath table, but you can make do in an ordinary shower or tub.

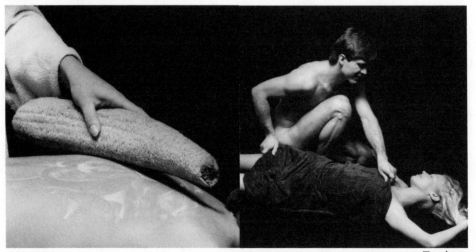

Loofah rub   *Towel wrap*

As Inkeles correctly points out, bathing the receiver is an "ideal introduction to the world of massage—you can lie back and do absolutely nothing but feel." It pays other dividends as well. "For most people the experience of being bathed becomes a dreamy sort of return to early childhood, the last time anyone touched or bathed their entire body," he adds. After the bath, you'll want to use the thickest, thirstiest, biggest towels you can to dry off your friend. Inkeles recommends heating the towels inside a large paper bag which you put in the oven for ten minutes at 200 degrees. The paper bag won't burn if it's not exposed to an open flame, *but be careful to watch it closely*.

Wrap the hot towels all over the receiver's body, starting at the feet and working up until he's completely covered. Pat and rub your towel-snuggled partner all over until dry. Let him luxuriate in the warm towels for a few minutes.

The most effective massage is given on a massage table. A good table allows you to move around easily and get the best leverage. The proper dimensions of the table are about six feet long, two feet wide, and thirty inches high. It should be topped by about two inches of foam padding, blankets, or other soft material. It must be sturdy, because a crash is the last thing your partner needs. There are portable massage tables (like the Astra model) which are steady enough to allow you to climb on and straddle the receiver. *New Massage* provides detailed plans for building your own combination massage/bath table.

In *Total Massage*, Jack Hofer offers this advice for making a simple massage table: "Take a piece of plywood or composition board that measures six feet by two feet. Using 1 by 4 inch pine, some nails and glue, build some sides and ends on your table. Then place four sturdy kitchen chairs with flat tops evenly under the table . . . The seats of the two end chairs should stick out from the table so you can sit down while working on the head or feet. Cover your table with foam, cloth, leather, vinyl, fur or whatever turns you on."

If you use the simplest route—the floor—supplement any carpeting with a large foam pad, several blankets, or a sleeping bag. Outdoors, a comfortable blanket over soft grass or sand should suffice. Whatever surface you use, leave a sheet immediately underneath your friend to catch any excess oil. Lubricating oil is essential for a full-body massage, unless you literally want to rub the receiver the wrong way.

Many of the movements designed to reach deep muscle tissue cause a good deal of friction. Oil will reduce this, while enhancing the sensual aspect of the experience.

When choosing an oil, don't bother with fancy-sounding, expensive commercial preparations. You can do just as well with easily-available, clear, light vegetable oils made of safflower, coconut and almond. These are natural products which can be taken internally, so they'll be safe for your skin. For a pleasant scent, add a small dab of scented oil like peppermint, wintergreen, cinnamon, or clove. Six drops of lemon extract per quart of vegetable oil will also do the job. When applying oil, place a few drops in your palms, and rub them together before laying your hands on the receiver. A couple of drops will do for most body parts, although hairy spots might take a bit more.

The best dispensers of oil are narrow-spouted plastic squeeze bottles used for condiments, shampoo or vaseline. These help prevent spills which can occur if you use a bowl. Inkeles suggests warming the oil in a glass bottle set in simmering hot water before beginning the massage. Transfer the oil to the squeeze bottle, and it'll stay warm for the length of the massage. After finishing some body area, the oil can be removed with a warm towel. At the Swedish Institute they also use alcohol, kept in a second squeeze bottle, to wash off the excess lubricant.

An electric vibrator can also accompany a massage, but it's best to use it *before* you start oiling the body. The best vibrators for a massage are the kind that strap onto the back of your hand. These allow your hand — as opposed to plastic or rubber — to make the actual body contact. The vibrator will stimulate circulation and titillate your partner, but the effect is much different from the softer rhythm of the massage. Therefore, use the vibrator as an appetizer, but not as part of the main massage. Use slow movements with the vibrator. It's especially effective on such extremities as the scalp, hands and feet, but it's alright to use the hand-strap variety all over the body. It is definitely a nice pre-massage warm-up.

Other accessories you may want to include are a large, billowy feather, a silk scarf, or a soft fur-piece.

The Setting A few words about atmosphere: The key concepts for the massage experience are *calm* and *mellow*. Anything that will create or enhance those states helps. Soft lighting, classical or other gentle background music, and delicate incense all add to the mood. Your partner will be very vulnerable, completely surrendering to your control, so it's important to keep him from jarring distractions. If someone other than you can't answer the telephone, then take it off the hook. Make sure children and pets are out of the way.

Protect your partner from drafts. While *you* may get sweated up, he can easily get a chill. The massage room should be about 75 degrees, but the surface the receiver rests on should be no cooler than 70 degrees. This may require an even warmer room. You can also keep him extra warm by covering with towels the body areas you're not working on.

Final preparations: Make sure your friend is in his birthday suit, sans jewelry, glasses or contact lenses. Have him lie down on his back on the massage surface, and join him in a few minutes of deep breathing. Make sure your hands are as warm as possible. Rub them together briskly, or press them under your armpits before starting. *Remember, the oil used will be rubbed on your palms, not directly on his body.*

To simplify matters, it's best to work with a set sequence in the full-body massage. This is the sequence we will be using here: torso, arms, hands, neck and head, front of legs, feet, back of legs, buttocks and back. Following a pre-arranged plan while learning keeps you from missing some spot. Once you become accustomed to it, you can invent your own sequence. The only rule to follow for sequences is *continuity*; don't jump around between widely separated areas. Avoid breaking contact whenever possible.

Basic Moves One of the secrets of a top-quality massage is variation in movement. The easy massage technique described above is much simpler, because it's not much different than just touching someone all over. The full-body massage accomplishes the same end, but in a much more sophisticated (and therapeutic) way. It will stimulate and soothe the skin — like easy massage — but it also tones the internal organs, tendons, joints, and other deeper places. The main difference between the two techniques is in the use of the hands. For easy massage, your hands basically worked as a single tool. For full-body massage, they function more like an entire toolbox. Here are the main moves you'll be using:

Effleurage: Pronounced "ef-flu-rahj," it is also known as stroking. The *Swedish Massage Work Book* (written by the director of the Swedish Institute, Dr. Sidney Zerinsky) describes effleurage as "gliding the hands with long, even strokes over the surface of the body." For circulatory reasons the strokes should usually go towards the direction of the heart. Effleurage is generally a light, stimulating movement, often used as an introductory manipulation for deeper work. It can be used all over the body, but works particularly well with the extremities.

Petrissage: Pronounced "pet-rah-sahj," this basically involves the kneading of muscle tissue. It includes pressing, squeezing, rolling and picking up muscles. It can be done with the hands, thumbs and — for small areas — with the thumb and forefinger. Petrissage boosts circulation to the muscles, and removes waste. It works on all muscular spots, as well as on soft, fleshy sections.

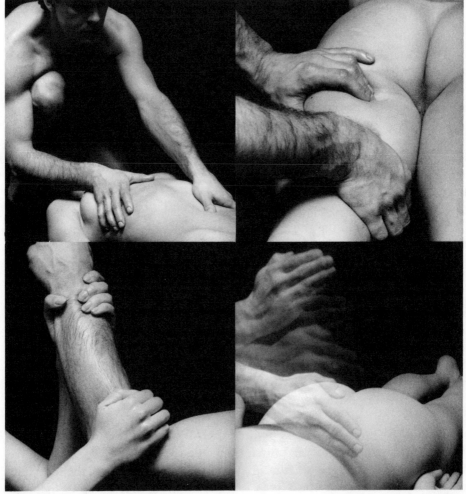

Effleurage; Petrissage; Friction; Tapotement.

Friction: This involves rapid, circular pressure over a particular area. On large surfaces like the back, it's done with the palms. Otherwise, the thumb or fingertips are used. Friction is good for treating internal tissues and joints.

Tapotement: Pronounced "ta-pot-mon." There are several variations of these short, quick blows with the hands or fingers: hacking (edge of the palm); slapping (flat of the hand); tapping (fingertips); cupping (cup-shaped hand); beating (edge of the fist). Tapotement stimulates nerves and muscles, and boosts circulation. It is used mainly on the back, and on the backs of the legs.

Vibration: Done with either the fingers or hands, this involves rapid shaking movements. Use back-and-forth, trembling pressure. Vibration quickens the nerves and stimulations the digestive organs.

The *Swedish Massage Work Book* recommends that the moves be applied in the following order: effleurage, petrissage, friction, tapotement, vibration. This is just a general formula, because not every move is appropriate for all body parts. The Swedish system also tends to *finish off* each area with some additional effleurage strokes.

Full-Body Massage Routine

The Torso: Begin in the abdomen area. Position yourself so both of your hands rest lightly on the center of the lower abdomen, thumbs next to each other and fingers fanning out towards the sides. Move your hands in gentle effleurage strokes towards the sides, then back to center. Like all movements in the routine, this will be repeated several times. Next, position yourself by the receiver's right side. Move your hands in a clockwise effleurage circle from the lower abdomen to just underneath the ribs. Then, knead the entire abdominal area in a clockwise circular motion (all abdomen movements are clockwise, following the path of the colon) one hand following the other. Gently lift and squeeze folds of flesh between thumb and fingers while kneading. Don't forget to include your partner's sides. Next, place your fingertips on the abdomen and begin a vibrating clockwise circle, shaking the area all over. To finish off, repeat several times the circular effleurage you started with.

We now focus on the chest. Position yourself behind the receiver's head, hands resting side-by-side just below the neck. Slide your hands down the chest to the sides at the ends of the rib cage. From there, move your hands up the sides, over the armpits and shoulders. Continue to slide them over the tops of the shoulders until they meet again in the original position. Repeat the entire movement several times. Next, remain behind your partner's head. Rest each hand on the top of each shoulder. Move both hands down the body in a sweeping stroke all the way to the pubic area, then back up again. This is one of the longest single strokes in the massage, so it should be repeated a number of times.

Reverse your posture so that you're now working from the opposite end of your friend. You can stay to one side or gently straddle him. Place each hand on one side of the rib cage, fingers pointed up. Move them towards the armpit, then across the chest (pectoral) muscles, then up to the neck. Repeat the entire movement a few times, as usual. Next, rest your right hand on the left chest muscle, fingers pointing up. Make a circular stroking motion up to the neck and over the left shoulder, down to the chest muscle again. After repeating this,

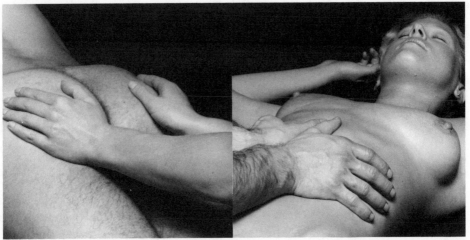

Vibration   *Starting position, abdomen*

duplicate it on the right side with your left hand.

    Next, place the fingertips of both hands on the right side of the neck. Knead across the back and side of the neck down to the shoulder. Repeat and duplicate on the other side. Next, place both hands in the middle of the chest, thumbs touching. Use oscillating friction movements down to the abdomen. Next, make your hands into a cup-shape, and gently tap over the bronchial area (this is for congestion). Finish the chest with some additional effleurage strokes.

The Arms: Begin with effleurage of the arm from wrist to shoulder. Hold the arm with one hand as you stroke with the other. Repeat as usual. In the next move, grasp the upper arm muscles with both hands. Twist your hands in opposite directions up and down the arm, fingers together. Next, place both hands underneath the arm, with thumbs on top. Knead the muscles up and down the arm with your thumbs in a circular motion. Next, bend your friend's arm over his chest. Support his forearm with one hand, while you place the palm of your other hand over his elbow. Push and twist in tiny circles against the elbow. The final moves are hacking up and down the arm with the edge of your hand, followed by tapping the same spots with a cupped hand. Finish up with some more effleurage strokes over the entire arm. Duplicate all movements on the other arm.

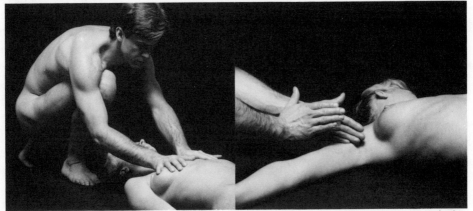

Length-wise body strokes   *Arm hacking*

The Hands: Start with thumb-stroking of the palm to warm it up. Steady your partner's hand on one side with one of your hands, while you stroke up and down the palm with your other hand. In the following move, sandwich his hand between both of yours. Then knead it with your palms, using a circular motion. Next, grasp the hand with both of your thumbs resting on its back. Use circular kneading motions on the palm side with your fingers, and back side with your thumbs. Work the bones of his hands this way. Now move up to the wrist joint and continue the circular kneading there. Then reverse your position, so that your thumbs rest against his palm. Press and knead with your thumb, especially around the heel. You can vary the pressure by switching to your knuckles here.

Now move on to his fingers. Support his hand in one of yours, while you press each finger and the thumb between *your* forefinger and thumb. Use a twisting, circular motion. Do the front and back of each finger, then the sides. Complete the hand work with some effleurage strokes on both sides of it. Duplicate all the movements on the other hand.

Hand kneading   *Neck strokes*

Head and Neck: First we'll focus on the neck. Position yourself behind your partner's head and place each hand on one side of his neck. Make small, circular moves with your fingers along the sides and back of the neck, from the base of the skull to the shoulders. Repeat as usual. Next, place both hands under the back of the neck, cradling the base of the skull. Let your thumbs hang free. Stroke back and forth on the neck, pressing up against the muscles. Keeping the same position, put the first few fingers right on the base of the skull, and begin vibrating motions. Use medium to firm pressure.

Remaining behind the head, place one hand over the top of your partner's head. Keep the other hand at the base of the skull to anchor it. Press down with the hand on top, and move it back and forth in circles. The scalp will move along with your hand. Go from the hairline on back, then gently lift the head to continue down the back of the neck. Next, knead the scalp all over with your fingers. Make small circles, covering the area a few times. *Don't use oil on the scalp*.

Before massaging the face, you may want to give your friend a special hot towel treatment. Jack Hofer suggests doing it this way: soak one towel in hot, *but not boiling*, water. Wring it out and mold it around your partner's face, except for the nose. Leave it on for 2-3 minutes, with your hands held lightly on the sides of the head. Replace the hot towel with one that's been soaking in cold, *but not icy*, water (wring it out first). Leave the cold towel on for about a minute, then tenderly rub the face with oil, lotion, or cream.

Gordon Inkeles doesn't bother with cold towels, but adds instead this twist—an herbal steam facial. His hot towel treatment uses a large pot and a half-dozen small towels. Fill the pot with sufficient water to cover the towels, then add a half-teaspoon of an ordinary spice-rack herb like thyme, sage, or lemon peel. When the towels are nice and warm, unroll one of them from the forehead down. In 2-3 minutes, replace it with another, and so on until you've used the whole batch.

After the exotic hot towel treatment, you're ready to work on your partner's face. Start with the forehead. Place one hand flat across his forehead, fingers facing to the side. Stroke straight down to the bridge of the nose, simultaneously replacing the first hand with your other hand. You'll have a smooth, hand-over-hand action, as one hand replaces the other. Next, place the

Hot towel treatment   *Head strokes*

front of the fingers of both hands on the forehead. Your fingers will face down towards the eyes. Make slow, circular movements from the middle of the forehead to the temples and sides of the head. Repeat as usual. From there, very gently place the pads of your forefingers on each eyelid (*but don't get any oil on them*). Very lightly stroke the eyelids with soft, circular motions. Now move your attention to the cheeks, placing one hand on each of them. Use oscillating movements on and around the cheeks. Next, bring the hands down below to the jaw and chin area. Continue the movements there.

The next focal point is the mouth. Massage the muscles there by circling the lips with your index or middle fingertip. Use medium pressure, moving in tiny circles. Gentle brushing of the lips themselves can be your next movement. Whatever small spaces of the face you haven't yet reached should be worked with circular pressure from a single fingertip. This includes spots like the sides of the nose. We now shift to the ears. Using the pads of the first few fingers, press all around the outer ear. Then use the same movement under the ear, with a bit more pressure on the bony parts beneath it. Repeat several times as usual, then complete the movements on the opposite ear.

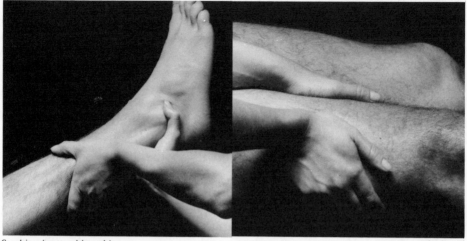

Stroking from ankle to hip   *Thigh rub*

The Legs: The first manipulation will be full-leg stroking. Start with your hands over one
    ankle, next to each other in a fan shape. Stroke all the way to the top of the thigh,
    using moderate pressure. Shift your hand position as the limb widens, making
    sure you cover the most area possible. Reverse the stroke back down to the ankle,
    and repeat several times.

    We'll now cover the sides of the leg. For the *outside*, grasp the ankle with one
    hand and raise the leg to a 45 degree angle (the right leg will be grasped with your
    right hand, the left leg with your left hand). Use your free hand to give smooth
    effleurage strokes from ankle to hip. For the *inside*, reverse your hand positions,
    so the hand that was free now grips the ankle. Use the other hand to stroke the
    inside from ankle to upper thigh.

    For the next move, return your hands to the ankle of the lowered leg.
    Using your thumbs, begin circular kneading motions up the entire leg. Next,
    bend the leg at the knee, with the foot flat on the massage surface. You can
    anchor the foot between your knees. Place your fingers under the calf muscle,
    with thumbs resting on the front side of the leg. Knead the top part of the calf
    with your thumbs, as you press the underside with your fingers. Work from
    the ankle to just beneath the knee.

    We'll now shift to the knee itself. Flex it slightly, using a pillow to support
    it if needed. Use thumb friction to work all the crevices around the kneecap.
    Circle around the knee in small, oscillating patterns. *However, don't press
    directly on the kneecap itself.* Next, knead the muscles of the leg from knee to
    hip. Use both hands, shaping your grip to the muscles themselves. Finish the
    thigh with some rolling manipulations. Place your hands opposite each other
    on the sides of the thighs. Move your hands back and forth in opposite
    directions, as you roll and wring the thigh muscles. Complete the leg with
    some more full-length strokes. Duplicate all movements on the other leg.

The Feet: To start, place both hands side-by-side over the ankle, perpendicular to it. Stroke
    down from ankle to toes, one hand following the other. In the next move, support
    the foot with your knee underneath. Place each of your hands on the side of each
    ankle, thumbs resting on top. Make circular kneading movements all around the
    ankle and foot. Next, keeping the same posture, knead up and down the top part
    of the foot with your thumbs, pressing the sole from beneath with your fingers.
    Make sure you get in between the small bones there.

Foot strokes   *Toe popping*

Now we'll shift to the toes. Knead each one up and down between thumb and forefinger. Pay particular attention to the big toe.

Next, wrap your hands around the foot so your thumbs meet on the sole while your fingers rest on the back. Knead the entire sole between thumbs and fingers. Use good pressure, as this is a heavily-padded spot. Repeat as usual, then do the other foot. After that, it's time for your partner to turn over on his stomach. From that position, rest one of his feet over your own leg. Return to the thumb-kneading of his sole. This time, concentrate on the arch (which is more accessible now), using firm pressure.

In the next move, bring the thumbs up over the arch to the Achilles tendon. Your fingers will rest under the ankle, securing it. Knead the tendon between both thumbs in a circular motion from the heel up to the calf. The final foot movement is knuckle-kneading. Anchor the flexed leg over your knees while you hold the foot in one hand. With your free hand, knead the sole and arch with the flat part of the knuckles. Dig in firmly, working your way up and down in circles. Duplicate all the movements on the other foot.

Backs of Legs and Buttocks: First, do effleurage strokes of the entire back of the leg from ankle to thigh. Your fingers should point straight up for this. Next, encircle the ankle with both hands, thumbs on top. Knead with your thumbs, working your way up in tiny oscillations. Get into the calf muscle thoroughly. From there, keep

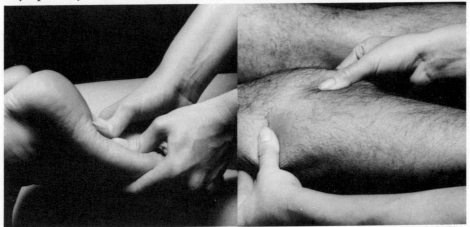

Sole kneading   *Calf plucking*

working up the whole thigh. The next move involves picking up the muscles from ankle to thigh.

Pluck the flesh between fingers and thumb, and pull the muscles slightly away from the bone. Next, place your hands on opposite sides of the upper thigh, and roll and wring the muscles the same way as you did the front of the thigh. Work from the knee to the hip. Lastly, begin percussion movements all over the leg. From ankle to hip, do hacking (edge of hand) and cupping (slaps with cup-shaped hand). For a final touch, try a few more full-length effleurage strokes. Complete the other leg.

Start the buttocks with some light strokes up and down with both hands. Then really dig in with some good kneading. You can either work the two cheeks simultaneously or one at a time. Make sure you cover every area, picking up generous hunks of flesh between fingers and thumb. Use deep pressure every-where *except on the spine itself*. Next, try some vibrating movements to shake the cheeks up good. Put your palms on top of the buttocks, and start rapid shaking movements up and down. This motion, as Hofer points out, will "vibrate the daylights out of all those tight-assed stored up feelings." Do it for about half a minute, then finish up with a few more light strokes.

The Back: Begin the back with effleurage strokes from the base of the spine to the head. Position your hands to each side of the spine, fingers pointed up, starting at the tailbone. Most back movements work *next* to the spine, not directly on it. However, it is okay to do some *light* strokes up and down the spine with the flat part of your hands. Just ease up on the pressure. Spinal contact directly affects the nerve branches throughout the body — fast movements stimulate, while slow and deeper rhythms sedate.

The next manipulation involves stretching the back muscles from the spine outwards. Position your palms opposite each other on each side of the spine. Fingers should face towards the side. Move up the back, stretching the muscles out from the spine. The following move entails kneading the muscles of the entire back: knead up and down the back, stretching, pressing and pulling the many muscles. Use your fingertips to work the hard-to-get spaces beneath the shoulder blades. Most of the back can easily handle deep pressure, if not outright crave it. Be thorough with the back, because it's an important area containing countless numbers of nerve connections.

In the next move, place your hands side-by-side on one side of your friend's back. Roll your hands from one side of the back to the other, each hand crisscross-ing the other. Start from the base of the spine and work to the top, using moderate pressure. The following move is spine vibration: place the first two fingers of one hand on each vertebra (on either side of the spine). Vibrate each vertebra, working from the top of the spine to the bottom.

We'll now focus on the neck and shoulders. Place your fingertips over the top of the shoulders, with thumbs sitting on the base of the neck. Work your fingertips and thumbs into the muscles, kneading all over. Move over the neck with the thumbs, using rhythmic, circular kneading. Get into all the tense muscles on the side of the neck. Before leaving the head area, you might want to give another short scalp massage.

Round out the back work with percussion movements. Cover the whole back with hacking, slapping and cupping movements. Finish off with a few

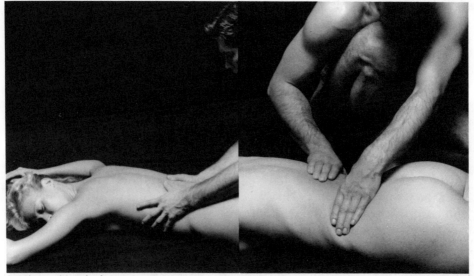

Starting position, back　　　　　　　　　　　　　　　　　　　　*Back rolling*

more strokes up and down the length of the back.

Alcohol and clean towels can be used to dry off any excess oil. You can also end the massage with a sensual treat. Drag the billowy feather, silk scarf, or fur-piece across your partner's back. Long, flowing hair will also work the same magic. Then let your partner lie there quietly for a while. Don't be over-anxious to raise him. You, too, can just lay back and derive pleasure from knowing what an exquisitely blissful experience you've given someone.

How long will this full-body massage take? If you repeat all the movements several times and add the special treats like hot towels, it could run almost two hours. A less elaborate version should take approximately 45 minutes. But time really shouldn't be a key consideration. The real goal is an extravagant tactile feast for your touch-partner. If you regularly treat each other to such magnificent massages a special closeness and understanding will grow between you, and you will feel great delight, not detachment, in the other's joy. Lovers who massage together may well stay together.

Group Massage In the world of massage, whatever one person can do, two or more can do better. Group massage is a unique sensual treat. It's wonderful to be worked on by a large number of tender, loving hands. Group massage movements are essentially the same as those used in the one-to-one full-body massage. The main difference is that many things are done at once.

Basically, the body is divided according to the number of available hands. Different people will work on different parts. For example, four people could each massage one limb. Since silence is still the rule, the roles should be coordinated before starting.

Hofer says that four-on-one is the practical limit for a comfortable massage, but Inkeles is a bit more expansive, setting nine-on-one as the absolute maximum and five-on-one as a happy medium (the odd person can always work the head area). Group massage can also be combined with the group sensitivity strokes described above. For example, the group can follow the massage by lifting and rocking the receiver.

Self-Massage Self-massage can't seriously compete with being catered to by another, but it does serve some useful purposes. It can be effectively combined with an exercise routine, or used as a quick stress-reliever. You can do many of the movements in bed, to help you off to a pleasant sleep.

As Hofer describes it, self-massage involves two postures: sitting up and lying down. In a seated position, you can do your legs, feet, neck, shoulders, arms, hands, face and head. Use whatever full-body massage techniques you like for these areas. To knead the shoulders, cross your arms at the chest so that each hand reaches the opposite shoulder.

Next, lie down and bend your knees, with feet flat on the floor or bed. Then do your chest and abdominal area. To finish with a flourish, Hofer makes this suggestion: "Stand up and massage your buttocks and slap everywhere on your body that you can comfortably reach. Then shake and vibrate your arms, hands, legs—let your entire body shake like a bowlful of Jell-O or like a skeleton rattling all its bones at once."

Self-massage, shoulders

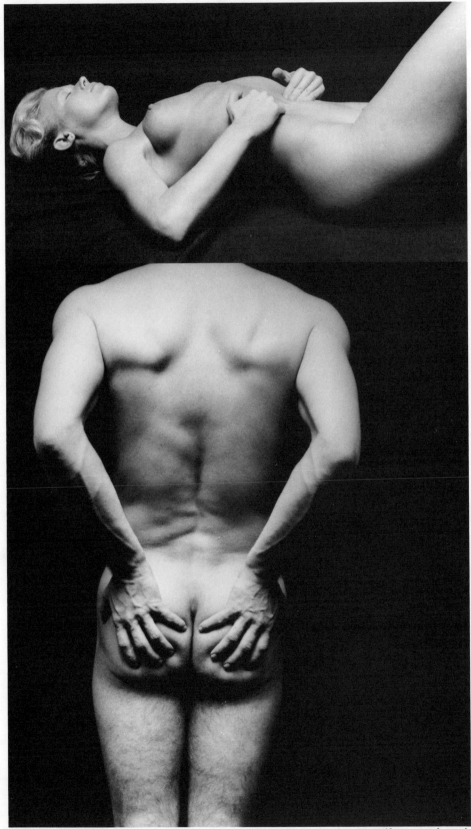

Self-massage, abdomen

Self-massage, buttocks

PART IV: TOUCH TECHNIQUES: HEALTH AND WELL-BEING

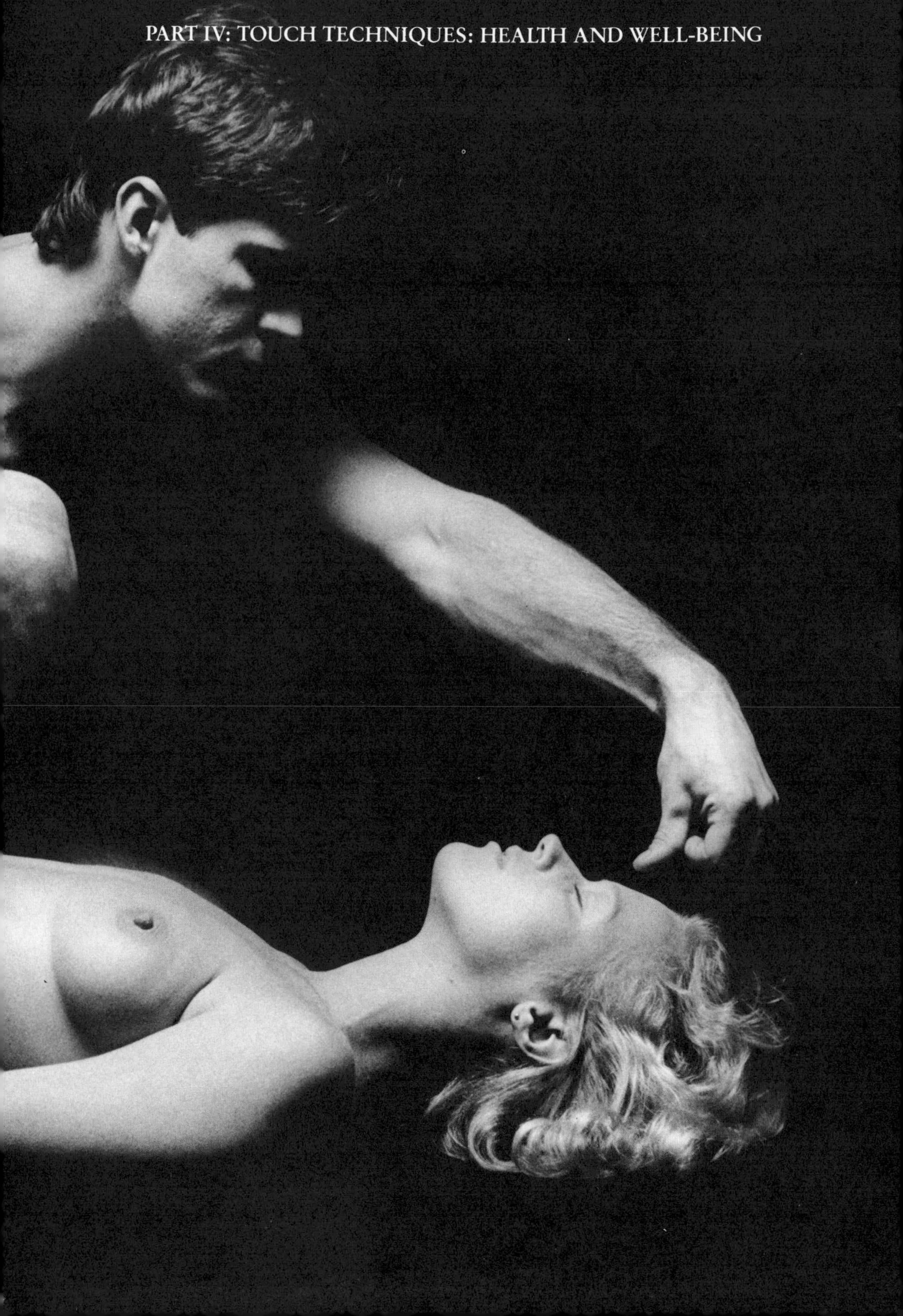

CHAPTER THIRTEEN: REFLEXOLOGY MASSAGE

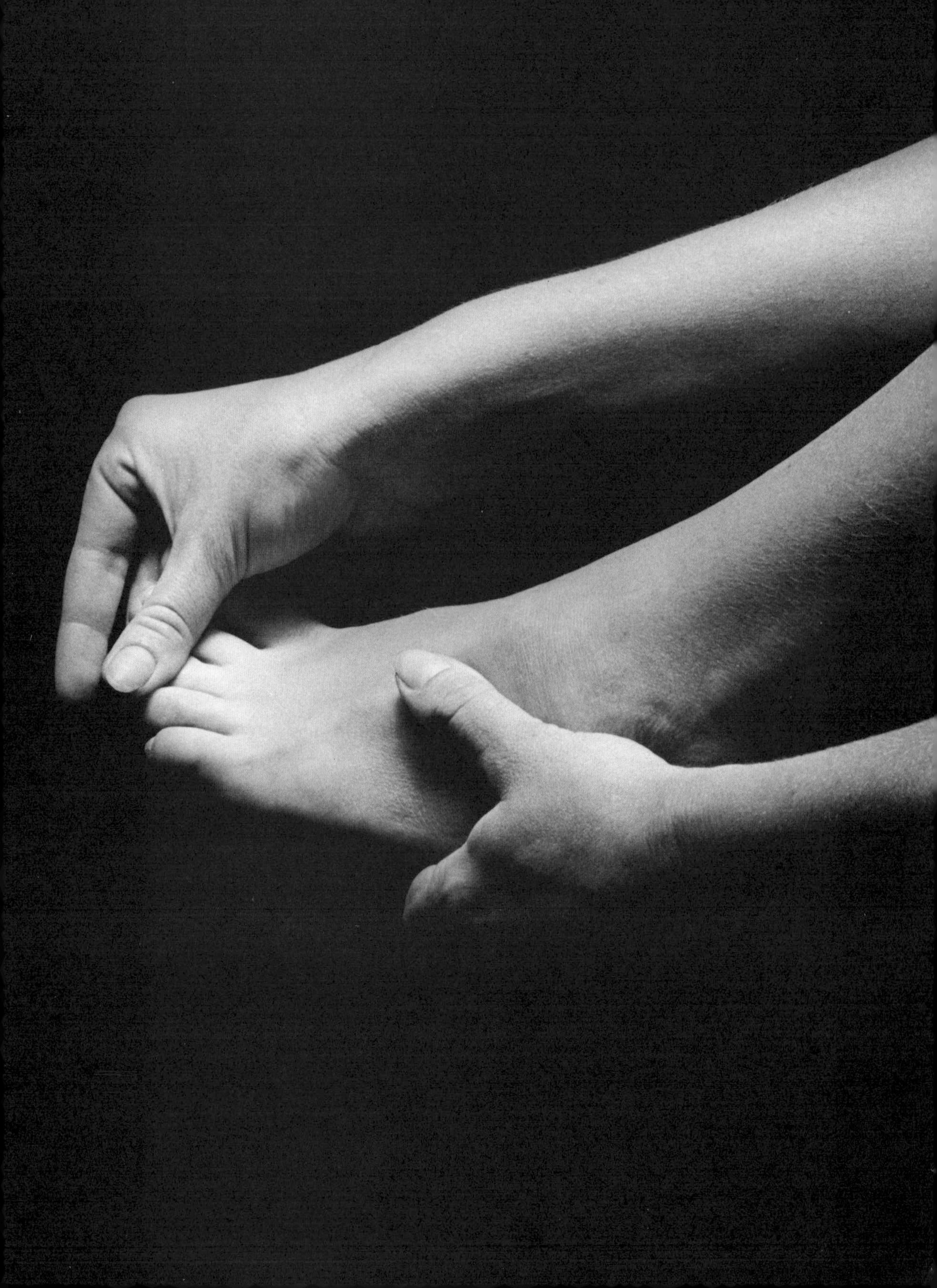

INTRODUCTION Part IV contains the most unusual, and perhaps controversial, materials in this entire book. The healing systems discussed here rely on a theory of health which is quite different from conventional Western medicine. To a greater or lesser extent, all the techniques employ a health model based on traditional Oriental medicine.

The basic idea behind this model is that *health depends upon being in harmony with the primal "life energy" of the universe*. This energy has been given various names, including *ki, prana* and *orgone*. The healthy person is supposed to be a "clear channel" for this energy — it constantly flows in and out of him through invisible pathways in the body. Sickness occurs when the energy flow becomes trapped or blocked in some body part. The sick person stops being a clear channel, for his energy circulation is obstructed.

The techniques for restoring health are all designed to remove the blocks and free up the energy flow again. Generally, they stimulate the invisible pathways ("meridians") to get the energy moving. Pressing on certain spots along a meridian ("pressure" or "reflex" point) stimulates all the internal organs along the meridian's path. The meridians are not based on conventional anatomy or physiology. A surgeon, for example, could not operate and locate them as in the case of veins or arteries. They may be based upon nerve reactions, but this hasn't been clearly established.

In his book, *New Massage*, Gordon Inkeles reports some interesting neuro-anatomy research regarding meridians. He states that Dr. Steven Colwell and his colleagues at the University of California have located tiny "maps" in the center of the brain which "may correspond quite precisely to acupuncture meridian charts." Perhaps, as Inkeles notes, "Western scientists have simply been looking in the wrong place to confirm the existence of the mysterious acupuncture meridians."

Wherever the meridians are located — if, indeed, they have a physical locale at all — their workings seem strange to Western minds. If your stomach aches, you normally react by taking something to directly soothe it. The act of pressing some distant spot on the body for relief elsewhere is obviously a rather different affair. Advocates of pressure-point healing often claim that their methods produce better results than do conventional techniques. In some cases, there are studies to back up their claims.

Based on personal experience, the healing techniques are generally simple and easy to learn, and can also be a lot of fun. Like all touch techniques, they stimulate and enliven the senses. If they make you *healthy*, as well as make you *feel good*, so much the better.

Try several of these techniques to see what they do for you. Take a playful, exploratory approach. Work alone on yourself or with your touch-partner. Of course, if you get good results don't react by clearing out your medicine chest or tearing up your doctor's card. A more cautious, wiser approach would be to integrate the new methods with conventional medicine, at least until you learn a lot more.

Professional instruction in almost all the techniques presented here has not brought any miracles, but occasionally some startling results have been noted which are difficult to explain within the usual medical framework. Let experience be your teacher, too.

The basic idea behind reflexology is that certain points on the body's surface are reflexes to internal organs. By pressing on a particular spot, one can stimulate and heal the corresponding organ. One can also diagnose the state of internal organs by examining reflex points; tenderness reveals the presence of disease or disorder.

According to Mildred Carter, author of several reflexology books and a practitioner for over twenty years, reflexology was introduced to Western culture by Dr. William Fitzgerald in 1913. He called it "zone therapy" and apparently derived it from ancient acupuncture techniques. *Reflexology divides the body into ten vertical zones*. Thus, each half of the body is separated into five analogous zones, numbered one to five. Zone 1 divides the body from the head down the center. An inch or two to the right of zone 1 lies zone 2 for the right side, while an inch or so to the left lies the corresponding zone 2 for that side. The remaining zones follow a similar pattern down the body. Branches of each zone run down the arms to the fingers, and down the legs to the toes.

These zones represent the invisible meridian lines through which the "life energies" are supposed to circulate. When some organ is diseased, the energy becomes obstructed along the meridian which runs through that organ. You can remove this blockage by pressing on a reflex point on the same meridian as the organ. As Ms. Carter explains in her book, *Hand Reflexology*, "Reflexology not only helps nature open up these channels when congested *but also sends a supply of 'prana,' the magnetic life force of the universe charging through these channels like healing shock waves* (her emphasis)." "When we press on all of the reflex buttons we are charging the whole body with a vibrating force of electric energy which stimulates nature into a speeded-up process of healing," she further explains.

Foot Reflexology The feet are undoubtably the most popular reflexology spots. The reflex theory holds that different parts of the foot correspond to different internal organs. Reflexes to organs on the left side of the body are found on the left foot, while the right foot contains reflexes to the right side organs. For example, reflexes to the heart are found on the left foot, and reflexes to the liver are on the right foot. Organs falling in the middle of the body can be stimulated on either foot. These include the intestines, stomach, and bladder.

The sole of the foot contains most of the key reflex points. Generally, the top, middle and bottom parts of the torso correspond to the top, middle and bottom parts of the sole. There are other significant reflex points on the top, inside, and outside of the foot. Apart from stimulating internal organs, reflex points also connect with key glands and nerves.

To get the most benefit from foot reflexology, cover both feet completely. At least at the beginning, be more concerned with massaging the entire foot than hitting particular points. If you reach an especially tender spot, however, pay more attention to it. If you're working on someone else, use your lap to rest the foot being worked on. If you're doing your own feet, sit down and rest the foot you're massaging on the opposite thigh. Foot massages are quite relaxing, so it's not a bad idea to do them before bed or a nap.

Technique Reflex points are located *under* the skin, not on it, so you must use deep pressure. Use deep, circular pressing movements. Be gentle, but press as deeply as you can without causing undue discomfort. If a spot is unusually sore, lighten up. The aim of all pressure point manipulation is to reach a sort of pleasure/pain response, a soothing ache which tells you you've been touched deep down. Just stroking the surface isn't enough, especially with the well-padded soles of the feet. *Dig in* with the balls of your thumb and fingers, but not, of course, to the point of pure agony.

We will be following Mildred Carter's suggested routine, as outlined in her book, *Helping Yourself With Foot Reflexology*. Start with your left foot. Secure it with your left hand while you press all over the big toe with the fingers and thumb of your right hand. The big toe is quite significant because it connects with the head. The center of the big toe corresponds to two important glands, the pituitary and the pineal, while the base represents the neck. The gland reflexes are quite deep, and might be better reached with the eraser end of a pencil. Cover all parts of the big toe, including the top (for the back of the neck), side (side of the neck), and base (tonsils and throat). After massaging, rotate the toe around (for loosening the neck) — then start on the big toe of the right foot. Big-toe massage helps to alleviate headaches, sore throats, and neck problems.

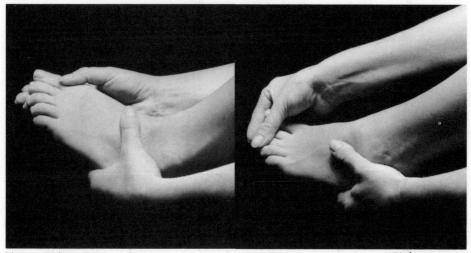

Big toe pressing   *Little toe pressing*

The next places to massage are the little toes. The bottom section of the two toes closest to the big toe contains reflexes to the eyes. The bottom section of the last two toes contain reflexes to the ears. Just below the ear and eye reflexes, slightly into the foot, will be found reflexes to the sinuses. Massage all these spots with your fingers, using a pressing, rolling motion. Spend extra time on any sore spots. Make sure you complete both sets of toes, since the left toes reflex to the left eye and ear, while the right toes reflex to the opposite side. Little toe massage helps with eye, ear, sinus problems and headaches.

The next reflex is the one to the spine. This runs along the right instep, starting just beneath the big toe. The bones which form the instep represent the vertebrae of the spine. Use the rolling, pressing motion of your thumb and finger to work this area from the big toe to the heel. Press in and under the bones. The top, middle and lower portions of the instep reflex to the corresponding sections of the spine. Unlike some of the gland reflexes, which should only be massaged for a few seconds, you can spend as much time as you want on reflexes to the back and other bony areas.

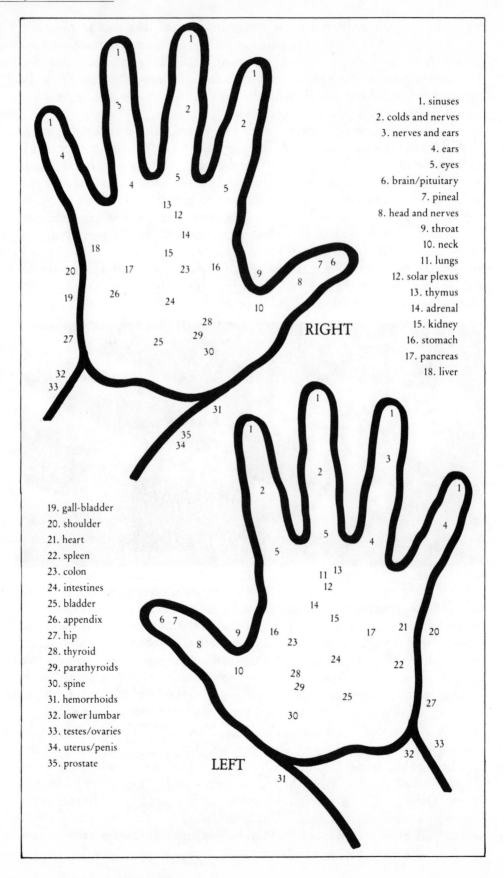

1. sinuses
2. colds and nerves
3. nerves and ears
4. ears
5. eyes
6. brain/pituitary
7. pineal
8. head and nerves
9. throat
10. neck
11. lungs
12. solar plexus
13. thymus
14. adrenal
15. kidney
16. stomach
17. pancreas
18. liver

RIGHT

19. gall-bladder
20. shoulder
21. heart
22. spleen
23. colon
24. intestines
25. bladder
26. appendix
27. hip
28. thyroid
29. parathyroids
30. spine
31. hemorrhoids
32. lower lumbar
33. testes/ovaries
34. uterus/penis
35. prostate

LEFT

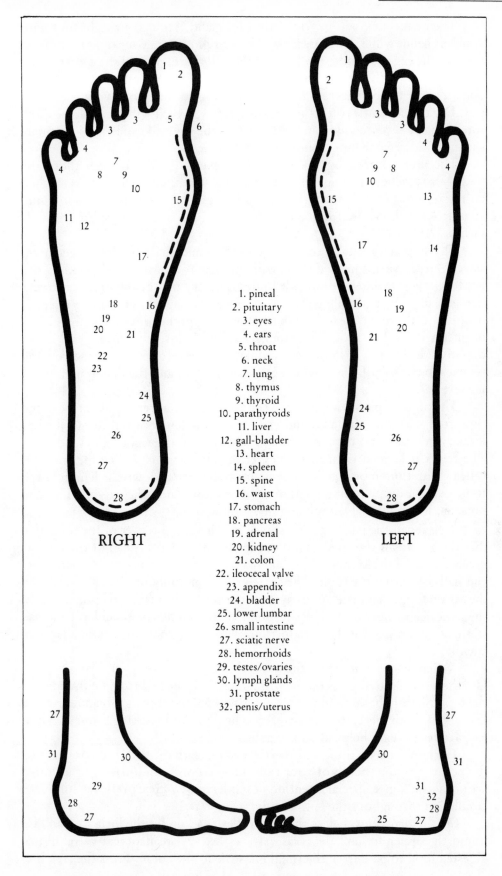

1. pineal
2. pituitary
3. eyes
4. ears
5. throat
6. neck
7. lung
8. thymus
9. thyroid
10. parathyroids
11. liver
12. gall-bladder
13. heart
14. spleen
15. spine
16. waist
17. stomach
18. pancreas
19. adrenal
20. kidney
21. colon
22. ileocecal valve
23. appendix
24. bladder
25. lower lumbar
26. small intestine
27. sciatic nerve
28. hemorrhoids
29. testes/ovaries
30. lymph glands
31. prostate
32. penis/uterus

RIGHT

LEFT

The next reflex will be to the thyroid gland. Hold your right foot with your left hand, while you massage with your right. Place your right thumb just beneath the big toe, under the bone of the pad there, toward the center of the sole. Start under the big toe and use a rolling, pressing thumb action under the bone. Follow the bone across to the outside edge of the foot. Work the whole pad. Dig in deep until you hit the thyroid reflex, which will probably be quite sore. Spend a few seconds there. Stimulating the thyroid can help to eliminate sluggish, lethargic feelings, and enhance alertness.

The area covered by the thyroid reflexes—the entire pad under the toes—also reaches the lungs and the bronchial tubes. The right lung is reached on the right foot while the left lung is massaged on the opposite foot. Lung stimulation helps loosen up congestion caused by colds and respiratory ailments.

The stomach reflexes are next. Use the knuckle of the forefinger here, since deep pressure is needed. Start with the right foot, using the knuckle of the left forefinger. Place your knuckle underneath the pad below the toes. Using the same pressing and rolling motion as above, massage the area under the pad, beneath the thyroid reflexes. Work from one side of the foot to the other. Then begin on the other foot. The stomach reflex helps with ulcers and other digestive disorders. Working the stomach reflex area will also act on the reflexes to the solar plexus (which cause relaxation) and the diaphragm (good for respiration).

The following position involves the kidneys and adrenal glands: work the right foot with the left hand, and vice versa. The left foot reflexes to the left kidney/adrenal, and the right foot corresponds to the right kidney/adrenal. The reflex is located right in the center of the foot, just below the stomach reflex. With thumb or knuckles, press from the center down the length of the foot, using a deep, rolling rhythm. Kidney stimulation helps you eliminate poisons, while adrenal stimulation invigorates you.

Reflexes to the pancreas and spleen also lie in the center of the foot. For these, work *across* the sole from one edge to the other, instead of the up and down action for the kidneys. The pancreas reflex runs clear across the left foot, and halfway across the right. The spleen reflex lies only on the left foot, near the outside edge. You'll get both by working across the center of both feet. Use the same deep-pressure, rolling rhythm with the thumbs or knuckles. Pancreas stimulation boosts insulin production, while spleen stimulation alleviates anemia.

We'll now concentrate on the liver reflex, which is only on the *right* foot. Use the left thumb to press on the pad underneath the little toe, on the edge of the foot. Slightly below and to the center will be found the gallbladder reflex. Use your thumb there, too. Liver reflexes help to eliminate sluggishness, while gallbladder reflexes help rid you of gallstones.

The same area on the *left* foot—the pad beneath the little toe—reflexes to the heart. Use the thumb of your right hand to press and massage the entire area. As usual, give special attention to tender spots. Heart reflexes stimulate circulation throughout the entire body.

On the *right* foot, below the liver reflex (pad under the little toe) will be found the appendix and ileocecal valve reflexes. The ileocecal valve allows digested food to pass from the small intestine to the colon. Press the area in a rolling motion with your left thumb. Stimulating these spots helps appendix problems and indigestion.

The next spot to be covered is the small intestine reflex. This covers virtually the entire pad of the heel. For deep pressure on this thickly-padded part, use a rolling motion with your knuckles. Start at the base of the heel and work your way up to the midline of the foot, moving across from the inner to the outer edge. This will help with gastric and other digestive disturbances. Work on both feet.

The following reflex connects with the colon. Start with the right foot. Use the thumb of the *right* hand to press just above the heel-pad, near the outer edge of the sole. Press up to the midline of the foot, then across the midline to the instep. Now switch to the left foot. Using your right hand again, start at the instep and move across the midline of the foot to the outer edge. Then move down from the midline to the top of the heel-pad. Search for tender spots all along the colon's path. Stimulation here helps eliminate waste materials and lessens constipation. Ms. Carter also claims that liver and colon reflexes are good for leg cramps and varicose veins.

Next in line is the reflex to the bladder. It's located close to the lower edge of the foot, near the instep. Press this area on both feet with your thumbs, using a rolling action. Use the right hand for the right foot, and left hand for the left foot. If you press deeper in the same spot, you'll reach the reflexes to the rectum, prostate and lower spine. Working the bladder reflex assists the urinary tract.

The outer edge of the heel contains reflexes to the hemorrhoids. In this manipulation, left hand massages right foot, and vice versa. Using thumb and forefinger, press along the bony edge, covering the entire heel. Do both feet. Right above the hemorrhoid reflex will be found the reflex to the sciatic nerve. This is off the center of the heel-pad, slightly lower and closer to the outer edge. This is a very deep reflex, and can best be reached with the eraser end of a pencil. If you find tender spots, use a deep rolling motion. Finish both feet.

You've just worked the end part of the sciatic nerve. To work other parts of it, leave the sole and move to the inner side of the ankle bone. Using your thumb, press above and in back of the ankle. This connects not only to the sciatic, but to all the reflexes to the lower parts of the body. Sciatic stimulation alleviates leg aches.

For the remainder of the foot massage, stay on top of the foot. The cord behind the ankle — the Achilles heel — contains reflexes to the sciatic, prostate, gonads and lumbar area of the back. Massage the cord between thumb and forefinger from the heel to the lower calf. Under the ankle bone on the outside

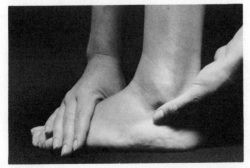

Ankle pressing

of each foot will be found reflexes to the ovaries and fallopian tubes in females and the testes in males. Press your thumb just below the ankle and above the heel bone.

On the inside of each foot under the ankle are reflexes to the female uteris and male prostate and penis. Use the right thumb on the right ankle and left thumb on the left ankle. Press with a rolling rhythm, moving back and forth across the whole area beneath the ankle. These last few moves stimulate the sex glands, relieve reproductive problems and, in the male, help with prostate congestion.

On the very top of the foot, above the ankle, are reflexes to the lymph glands. Using right thumb for right foot and left thumb for left foot, massage the entire area above the ankle, from the inside ankle bone to the outside ankle bone. Stimulating the lymph glands removes poisons which accumulate in the tissues.

That completes the foot massage routine, in a series of moves similar to those outlined by Mildred Carter. You may wonder why there is so much switching from foot to foot, when it would be simpler to finish one foot, then move on to the next. Ms. Carter's strategy seems to be to stimulate the corresponding area in each foot at the same time. That way, you concentrate your efforts on the same reflex points, rather than skipping around.

That makes sense, but it also complicates the massage quite a bit. In the beginning, when you're learning the reflex points, it pays to stick to her system. But once you become more familiar with it, there should be no great harm in doing one foot at a time. That would save time and energy, making the massage an easier, more enjoyable experience. Trade regular foot massages with your touch-partner; you both should find them to be soothing, relaxing and rejuvenating.

Hand Massage Hand reflexology is based upon the same principles as foot reflexology. Once again, the idea is to stimulate internal tissues by pressing reflex points. The reflexes on the hands parallel those on the feet, so you get the same benefits from both techniques. They can be used to complement each other, especially if you have a problem you really want to concentrate on.

For example, if you find that foot reflexes doesn't fully remedy some problem, you can switch to hand massage. Or you can experiment with the twin techniques and stick with the one that suits you best. One point to keep in mind: hand reflexes can be worked almost anytime, while you're much more restricted with foot massage. You can't just take your shoes and socks off anywhere. Not only are hands more easily available, but people commonly play with them in public, making hand massage even less conspicuous. In fact, Mildred Carter suggests that many popular forms of hand-fiddling (like stringing prayer beads) may reflect the unconscious use of reflexology.

While hand and foot reflexes connect to the same spots, the locations are not completely analogous, and the rules differ slightly. Here again, the Mildred Carter system, as outlined in her book, *Hand Reflexology*.

Technique You'll be using thumbs or fingers to press into the reflex points. Use pressure which is deep enough to reach the bones or muscles below, not just the skin surface. The basic motion will be a pressing, rolling action designed to search for sore spots. Tender areas signal trouble in corresponding internal organs, so pay special attention to them.

We'll start with the thumbs, which reflect to the head area. The right thumb connects with the right side of the head, while the left thumb does the same for the opposite side. The following movements assume you are working on your own hands. If you're working with a partner, the giver will support the receiver's hand in one hand, and do the massage with the other hand.

Massage the tip of the thumb by rolling the nail of the opposite thumb over it. Then grasp the thumb between the other hand's thumb and forefinger, and press deep into the thumb-pad with the edge of the other thumb. For deeper pressure, you can use the eraser end of a pencil. Work the entire pad with a pressing, rolling rhythm. Next, massage the back of the thumb (nail side) all the way down, once again using the edge of your other thumb. Then do the sides of the thumb. Finish by massaging the front part under the pad all the way down. Switch to the other thumb. The thumb reflexes to the brain, nerves, pineal and pituitary glands. The area around the base of the thumb connects with the throat and neck.

Now move to the fingers, starting with the index finger. Do the tip by rolling the opposite-hand thumb-nail over it. This hits sinus reflexes. Next, grasp the finger between opposite-hand thumb and forefinger, and press it all the way down, front, back and sides. When you reach the web between the index and middle finger, press it also. Repeat the same manipulations on the middle finger, which also reflexes to the sinuses. Next, massage the final two fingers in the same manner, making sure you get the webs in between all of them. The last two fingers contain reflexes to the ears and outer edge of the body. Finish the fingers on the other hand. Finger massage is good for colds and head problems.

The next area is the section immediately at the base of each finger. These spots reflex to the eyes and ears. Search for soreness with a pressing action of the opposite-hand thumb. Do both hands. Next, use the same pressing movement on the pad beneath the base of the fingers. This connects to the lungs, and is good for congestion and respiratory ailments. As usual, complete both hands.

Next, massage the soft spot in front of the thumb, slightly below the web. This reflexes to the stomach. Finish the other hand, then move to the center of the palm. Many reflexes overlap there, including the solar plexus, adrenal, kidney, pancreas, and thymus gland. Press into this spot with your thumb, as you press the back of the hand with your fingers. Do the other hand, then shift to the inside edge of the large bone, in front of and beneath the thumb. This links to the thyroid gland, which speeds up the entire system. Press deeply along the whole padded area, then do the same spot on the other hand.

Next, massage the pad under the little finger down to the midpoint of the palm. On the right hand, this area reflexes to the liver, gallbladder and appendix. On the left hand, the same spot covers the heart and spleen. On both hands, the outer edge of this area links to the shoulders, left hand for left shoulder and right hand for right shoulder.

Now continue on down the outer edge, from the midpoint of the hand to the base. In both hands, this area reflexes to the hips, each hand covering the hip on its side. From there, move away from the edge towards the lower central part of the palms. Press and massage this area up, down and all around with your thumb. On both hands, this spot connects with the intestines, colon and bladder. Next, thoroughly knead and press the web located between thumb and fingers. Grip this spot between the thumb and fingers of the opposite hand. In both hands, this links to the throat, neck, spine, stomach and other organs.

Reflexes to the spine are located in the back of the hand. They run down the bone which connects at the base of the index finger, and continues straight down the back of the hand to the wrist, below the thumb. Massage this line with the edge of your fingers. Do both hands. Next, move your attention to the wrist. On both hands, the bony edge of the wrist (front side) contains hemorrhoidal reflexes. Press the entire bone with the ball of your thumb.

Now drop slightly below the wrist, to the outer side under the pinky. Pressing this spot hits reflexes to the lower lumbar area of the back, the testes (in males) and ovaries (in females). Moving to the center of the wrist, we find additional reflexes to the lower lumbar. Finally, on the other side of the wrist (under the thumb) are connections with the prostate and penis (males) and uterus (females). For all these wrist points, don't forget to work both hands.

That is the basic hand massage routine. As for which points help which ailments, that information has already been given in the foot massage section. For example, if a foot massage reflex helps with headaches, the analogous hand massage reflex will do the same. As can be seen, hand reflexology is faster and simpler than the foot variety. Not only are the hands more accessible, but there's less territory to cover. Nonetheless, the greater surface of the foot means that the reflexes are less cramped, so you may be able to do a more thorough job there.

Both types of massage will stimulate and relax you. Overall, some find foot massage more gratifying, possibly because the hands are already involved in so many activities. Not only are the feet more neglected, but they're also under wraps most of the time. This lends naked footwork a certain private, intimate air. I particularly like to share foot massage with touch-partners. In contrast, the simple, fast hand massage is a quick lift you can give yourself any time. Which doesn't mean you shouldn't share hand massages with someone else. Whatever suits you best.

Body Reflexology Reflex points are by no means limited to the hands and feet. Mildred Carter says that there are more than 800 body points which can be tapped — among them, "ear reflexology" and "tongue reflexology." In another vein, a technique called "iridology" diagnoses conditions throughout the body by examining sections of the eye's iris.

Reflexologist William H. Oliver presents a full-body reflexology routine in his interesting book, *New Body Reflexology*. Oliver outlines reflexive manipulations for the entire body, and he feels this technique is superior to foot reflexology. He may well be right, and his informative, illustrated manual is recommended for anyone interested in learning more about the subject.

Body reflexology is not discussed in detail here, because it has many similarities to the Shiatsu methods described in the next chapter. (Both employ pressure point manipulations throughout the body to reach internal organs.) So while Westerners might be more comfortable with Oliver's presentation, some readers may prefer to learn the widely-practiced Shiatsu system.

CHAPTER FOURTEEN: SHIATSU MASSAGE

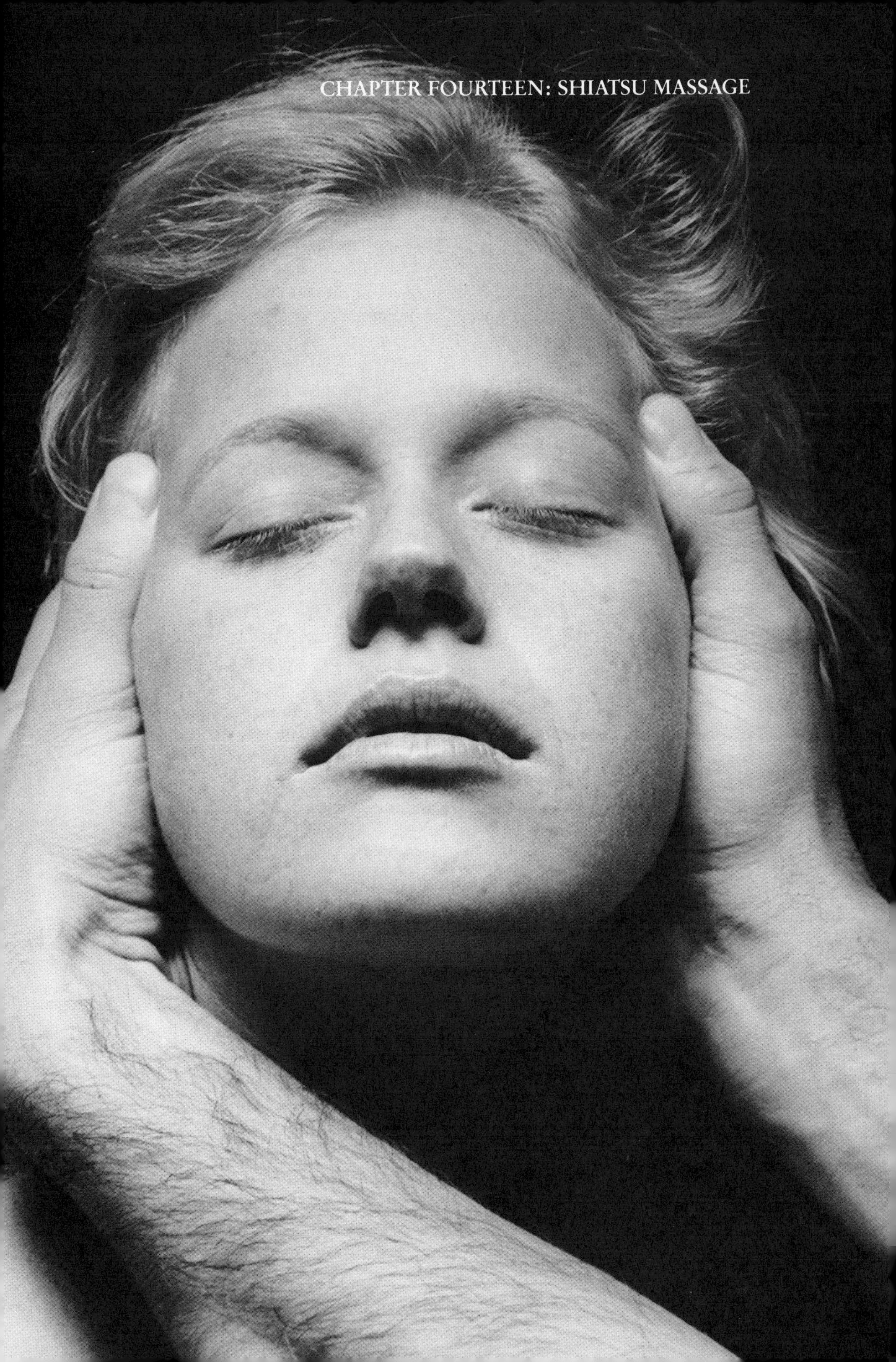

Shiatsu is Japanese for "finger pressure." It is also known as "acupressure." The technique derives from the same ancient Chinese medicinal principles which produced acupuncture. Basically, Shiatsu substitutes the hands for the acupuncture needles.

Shiatsu was introduced to Japan in the 6th century by Buddhist monks from China. Its present popularity, however, stems from the revival efforts of Tokujuro Namikoshi. He founded the Nippon Shiatsu School in 1930, and it wasn't long before his ideas began spreading west. For example, one of his students, Wataru Ohashi, founded the Shiatsu Education Center of America in 1975. The New York-based non-profit organization sponsors Shiatsu instruction in cities throughout the U.S. and Europe.

Like reflexology, Shiatsu divides the body into invisible pathways called meridians. These pathways link internal organs with points on the skin surface. Disease or disorder in an organ shows up by soreness in its pressure points. According to traditional Chinese medicine, disease results from a blockage in the vital life energy ("ki"), which is supposed to freely flow through each meridian. Stimulating the pressure points to an afflicted organ can remove this obstruction.

The main difference between reflexology and Shiatsu theory is the locations of the meridians. While reflexology divides the body into ten even vertical zones, Shiatsu establishes *twelve* main meridians (there are other subsidiary systems) which are anything but even. While most of the Shiatsu meridians orginate or end in the fingers or toes, they course throughout the body in confusing, zigzig patterns. By comparison, the reflexology meridian system is a model of simplicity.

Just how do these meridians work? Dr. Felix Mann, a physician who uses acupuncture in his London practice, feels that there is some as yet undiscovered nerve connection. In his book, *Acupuncture*, Mann says that "although the meridians do not exist as such, they illustrate in an almost abstract manner the presumed neural pathways which are as yet unknown . . . (they) might even be compared to the meridians of geography: imaginary but useful."

While Mann doesn't consider the meridians a physical reality (an opinion by no means shared by most practitioners), he still thinks acupuncture is an effective technique. "I practice acupuncture exclusively about 90% of my time," he notes in his book. "I would not do so if I did not achieve *better* results than in practicing Western medicine in the appropriate type of disease or dysfunction." To back up his claims, Mann provides statistics about thousands of successful treatments worldwide.

Of course, acupuncture successes don't necessarily translate directly to Shiatsu. Some observers argue that it's impossible for an area as large as the fingers to stimulate like a needle can. These critics usually consider the pressure points to be tiny and exact. However, others feel that the points are not nearly so small, or that so many cluster together that you're bound to hit one. Needle-pressure does seem to produce quicker, more dramatic results, so it may well be more effective than finger pressure. Nonetheless, many people are turned off by needles. And they penetrate the skin, so there's also the possibility of infection. Obviously, acupuncture administration requires more sophistication. Shiatsu is a much simpler, non-intimidating technique. Whatever — if anything — it loses in effectiveness, it makes up in ease of application.

Simplified Shiatsu As mentioned, the Shiatsu meridian system is complex and confusing. If you want some idea of this, see the meridian line descriptions in the "Touch For Health" section of the next chapter. Most manuals on Oriental massage are full of drawings illustrating these perplexing patterns. Nevertheless, they are not absolutely essential for effective Shiatsu. A second source of confusion is the seemingly endless series of pressure points. Dr. Mann says that he has seen descriptions of over *1,000* such points in Chinese literature ("Some books mention so many acupuncture points that one wonders if there is any normal skin left," he notes in *Acupuncture*).

Shiatsu Education Center materials focus on about 360 points, still a large number to keep in mind. Most technique manuals are crammed with diagrams of people with dots all over their bodies. Between these diagrams and the meridian line drawings, it sometimes seems as though one is reading a child's connect-the-dots coloring book.

Shiatsu instruction sessions led by Dyanna Yee and her husband, David Wong Yee, include a thorough Shiatsu massage routine which does not involve isolating a lot of pressure points. The couple received training at the Shiatsu Education Center and other places. By systematically covering the body, they reach most of the significant points. Instead of focusing on specific points, the following general routine is designed to hit the key spots. This will enable you and your partner to concentrate on each other, rather than on a bunch of lines and dots.

Basic Technique Correct pressure is the key to effective Shiatsu. *Remember, this is not Swedish-style massage with its sensual strokes, but a system whose benefits come from* deep *manipulation.* The pressure should be hard enough to evoke a pleasure/pain response. In most cases, you will apply this pressure with the balls of your thumbs. However, you should not usually rely on finger pressure alone; add to it the weight of your body. When possible, *lean in* with your body weight for a deep, steady pressure.

Naturally, you don't want to cause any sharp pain. Be guided by signs of obvious discomfort. Points which are particularly tender may require reduced pressure. At least in the beginning, it's not a bad idea to have your partner tell you when it really hurts. But you also want to avoid the common mistake of being over-gentle. Only good, firm manipulations produce beneficial results. One way to get the feel of the correct pressure is to place a thumb on a bathroom scale. Lean until the scale registers 12 to 15 pounds (24 to 30 with both thumbs). This is a good general pressure for most body parts.

You can practice this pressure on your own body. Try it on your thighs, forearms and head. A good, sensitive spot on which to test it is the bone which surrounds the eye sockets. Using one thumb for each eye, press the bone all around the eyes, especially at the temples. This should give you a clear idea of the pleasure/pain response you're looking for. By the way, this movement is beneficial for headaches and hangovers. When you do a full Shiatsu routine, you repeat each move two or three times. If you get a severe pain reaction somewhere, return to it with reduced pressure. Each spot you work should be pressed from three to five seconds in duration.

There's no special equipment needed for Shiatsu. You can use a massage table, mat, or carpeted floor. (A bed creates difficulty in balancing.) Oil is not necessary — nor advisable — for Shiatsu, nor is removal of clothes. However, while you can rough it, there's no reason not to enhance the experience. Soft music, subdued lighting, and a relaxed setting are all nice. Nudity for both partners increases intimacy and closeness. Another mood-enhancer is some deep breathing right before beginning.

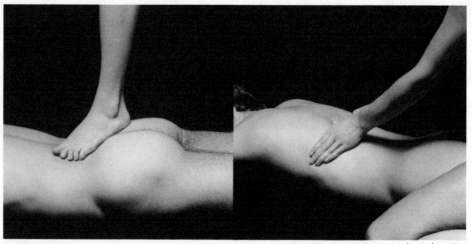

Back warmup   *Pressure along the spine*

Simplified Shiatsu Cycle Have your partner lie on his stomach, arms at his side, chin relaxed to one side. The back is a good place to start, since most manipulations there result in relaxation. If you want to warm up your friend a bit, put one of your feet squarely on the base of his spine. Bend your leg at the knee, and rest about twenty percent of your weight on the extended foot. Rock his spine gently back and forth with your foot. Gradually move this rocking motion up the spine to around the shoulder blades (it gets less sturdy above that point). This will loosen up and relax your friend.

   The first Shiatsu move is finger pressure along the spine. Starting at the top of the back, place your thumbs along the spine, less than an inch to each side. Press down the length of the spine, one inch at a time. Keep your arms straight, and lean your body weight into them when pressing. Don't press directly *on* the spine, but to the *side* of it. Breathe out when you press and in when you let go. As with all the movements to follow, repeat this two or three times, holding the pressure for three to five seconds on each point.

   Next, find the place where the hip bone connects with the spine. Place your thumbs along the spine at this point (one on each side) and press out along the hipbone, moving an inch at a time. Repeat the move several times, then return your thumbs to the hip bone/spinal connection. This time, press down and out in a V-shape pattern until you reach the sides of his body (the last points should be right about where the buttocks end). In both these moves, you'll be using both thumbs at the same time, one moving to one side and the other moving to the opposite side. In the next move, place each thumb about one inch below the top of each shoulder, not too far from the spine. Press out toward the end of the shoulders an inch at a time, then continue down along the outer edge of the shoulder blades. Stop at the bottom of the shoulder blades.

   Next, place along the inner edge of the shoulder blade on each side moving an inch at a time as usual. Next, starting with the thumbs on the outer edge of the shoulder blades, press in towards the spine in an inverted V-shaped pattern, reaching the sides of the spine around midback. Shiatsu on the back stimulates the entire body, and is particularly useful for back and nerve problems. We'll now focus on the neck and shoulders. Use less pressure on the neck. Place the thumb and forefinger of one hand on each side of the neck vertebrae, starting at the base of the skull. Use the other hand to balance. Press down the neck vertebrae one inch at a time until you reach the spine.

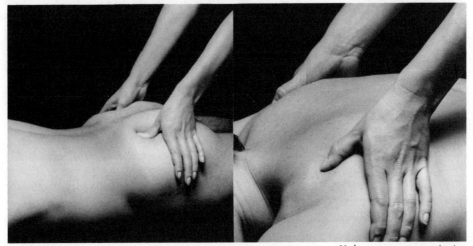

Hip bone pressure   *V-shape pattern toward spine*

Next, move your thumb to the side of the neck, directly under the ears. Work one side at a time here, moving down inch by inch. From the side of the neck, continue to press along the top of the shoulders. Do both sides. From there, try some pressing movements on the back of the shoulder muscle and in the shoulder blade area. New we'll shift to the back of the head. Press along the base of the skull in a straight line, from side to side. Next, place one thumb in the center of the back of the head, above the base. Press in a straight line all the way up the head. Continue over the top of the head to the hairline, or as far as you can comfortably reach.

Next, move one thumb about an inch to the *right* of the center line you just worked on. Press the head in a straight upward line as before, continuing over the top. Then move about an inch to the *left* of the center line, and repeat the

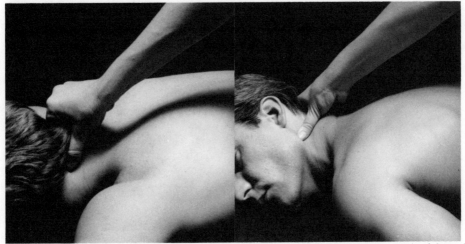

Base of the skull   *Side of the neck*

prior movement there. Next, move *two* inches to the right of the center line and press upward in a line from there. Do the same thing two inches to the left of center. Maintain these pressure lines up the head until you reach the sides.

It's probably wiser to work the head with one hand at a time. You can then use the other hand to cradle the head or to balance yourself. You may also find

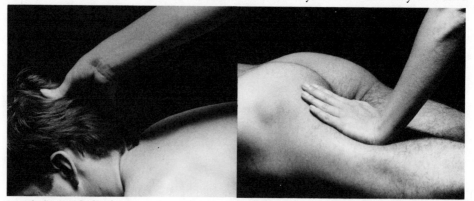

Straight line up the head   *Palm pressing back of the thigh*

it easier to work the head with your partner sitting up, which minimizes the possibility of his or her circulation being cut off. In any event, avoid excess pressure here. Shiatsu on the head and neck loosens tight shoulders and alleviates headaches and tensions.

The next spots we'll work on are the backs of the legs. Massage here is good for sciatic nerve pain and muscle cramps. Start at the top of the thigh, just under the buttocks. The area to be covered runs in a straight line down the center of the leg all the way to the heel. You can press down inch by inch using either thumbs or palms. See which is most comfortable for you and your partner. Don't press directly on the center of the knee. Some tension spots, like the calves, may also be too sensitive at first for hard thumb pressure, so you may want to use the palms instead. Make sure you complete both legs.

Now have your partner turn over and lie on his back. Reflexologists would maintain that the feet are literally covered with sensitive pressure points. Press your thumbs up and down the entire length of each sole, leaving no part uncovered. Then use fingertip pressure to hit all the points on the top and sides of the foot, around the heel, ankle, etc. While Shiatsu manuals don't make much of the toes, it pays to give them some attention, too. You can press them, wiggle them, or pull on them. Do both feet.

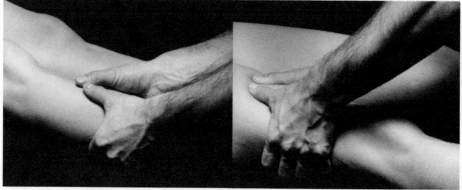

Calf pressure   *Thigh pressure*

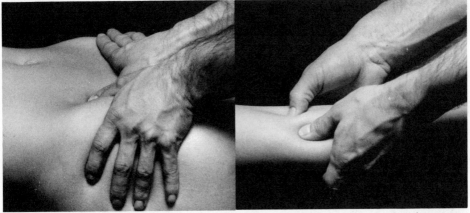

Pressing over pubic area   *Arm pressure*

Next, wrap both of your hands around the lowest part of one of your partner's calves. Your thumbs will meet on top of the calf, while your fingers will be underneath. You'll move up the calf to the knee, pressing the muscle between your fingers, while your thumbs press on top. Be careful here, because this is usually a really tense spot. Use as much pressure on the calf as your friend can handle, but don't drive him up the wall. If he's very sensitive, go over this movement repeatedly, using less pressure. When you reach the knee area, don't press on the cap directly. Instead, use some lighter massage movements around it.

The upper thighs are also loaded with pressure points. You can cover the inside, outside and front of the thigh in parallel lines, pressing up bit by bit with your thumbs. Start above the knee and move up to the hip. Alternately, you can press the entire area with your palms. Complete the full calf and thigh routine on both legs. Next, concentrate on the torso. The abdomen contains numerous pressure points to the intestines, stomach and other internal organs. Digestive disorders like ulcers and constipation can be treated here, as can menstrual troubles.

You can press the area between the navel and pubis with your thumbs, but the fleshy part of the abdomen is probably best worked with the palms or flats of the fingers. Pressing your thumbs in a horizontal line right above the pubic hair is good for the bladder and prostate. In the abdomen area press all over with firm, but not overwhelming, pressure from your palms or the flat part of your fingers. Lessen the pressure even more as you move up to the ribcage area. Continue the palm or finger pressure over the chest muscles and clavicle bone, on to the shoulders. This chest work is good for colds and congestion. For the shoulder muscles themselves, return to deeper thumb pressure, doing the front of the muscle, then the sides.

You're now in position to do the arms. They are also crisscrossed with pressure points, so there's not much use trying to isolate any of them. The best approach is to cover the entire arm with thumb and finger pressure. You can support the arm on top of your leg if you're using the floor, and press the front, back and sides, working in parallel lines from shoulder to wrist. Shiatsu here helps with tight, sore muscles. Make sure you also press around the wrist itself. The next spot, the hand, is loaded with pressure points, just as reflexologists would claim. Cradle your friend's hand between yours, with your thumbs resting on his palm. Press deeply with your thumbs all over his palm and heel, while your fingers work on top.

Then reverse your position so that your thumbs rest on the back of his hand. Thumb-press all over that spot. Finish the hands with some finger massage: pressing, rotating, pulling. Both hand and finger massage are especially good for arthritis. Make sure you work on both arms. The last body parts to do are the neck and face. You may prefer to work these spots when the person is sitting up. If you decide to do that, you can save all the neck and head work, front and back, for the beginning or end of the masasage, when you can do it all at once.

In the throat area, you'll want to place the thumb and forefinger of one hand on either side of the windpipe. Press *around* the windpipe —*not on it* — up and down the throat. Obviously, you won't be pressing too hard here. Next, place the fingertips of both hands along the muscles which lie on the sides of the neck, and press this area up and down, using *moderate* pressure. These movements are useful for sore throats, laryngitis, and other similar problems.

Next, press with your thumbs from chin to jaw, along the lower jaw bone. Then move up to the hollows of the cheek bones, and press in with one thumb on each side. Next, press the cheek bone area under the eyes, down as far as the bottom of the nose, and out as far as the sides of the face. Then pinch the bridge of the nose between thumb and forefinger. Next, press along the bone on top of the eye sockets, starting at the bridge of the nose and working all the way to the ears. Follow that with a similar move on the bone *under* the eye sockets. Do both sides at the same time. When you reach the temples, exert a few solid thumb presses toward the center of the head, pressing both temples simultaneously.

Next, move one of your thumbs right between the eyebrows. Press up the center of the forehead till you reach the hairline. Now place one thumb on each temple, at a point immediately in front of the tops of the ears. Press up the side of the head inch by inch until your thumbs meet at the top. Next, complete any areas on the top of the head which you couldn't reach when you were working from the back. Facial and head Shiatsu are good for sinus problems, eye strain, facial cramps, headaches and hangovers.

That's the end of the complete Shiatsu routine. It is a good idea for both partners to rest for awhile. The massage can be tiresome for the giver, but the receiver can also be a bit drained from the constant pleasure/pain stimulation. This takes some getting used to, but it is worth the effort. Shiatsu gives a unique sort of deep-seated gratification, as though your whole body has been embraced in a strong, yet loving bearhug. It relaxes as well as rejuvenates. Practice by swapping full or partial Shiatsu massages with your touch-partner. With regular use, you'll undoubtedly feel more energetic and alert, and you may suffer fewer chronic, nagging conditions. You can also experiment with combinations of Shiatsu and Swedish-style movements, creating your own massage smorgasbord.

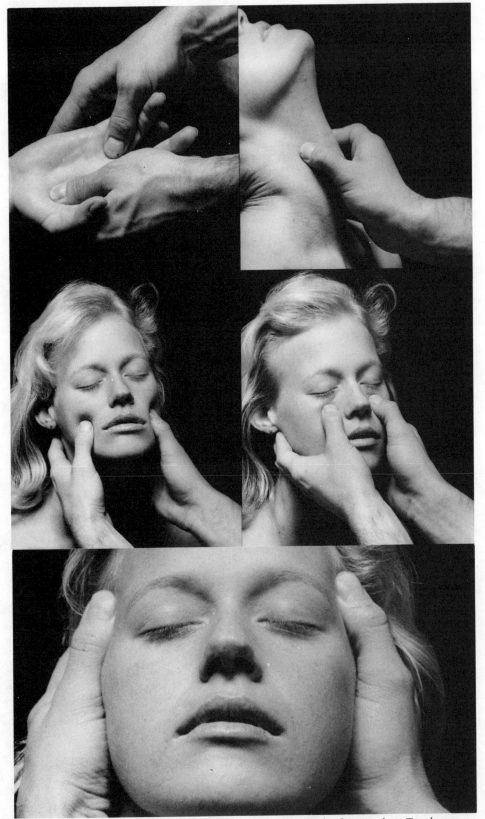

Palm kneading; Pressing around windpipe; Cheek bone pressure; Under the eye sockets; Temple pressure.

"Bodywork" is a general term which refers to a number of full-body touch therapies. This is a fast-growing area, with new techniques or variations popping up all the time. People use bodywork for all sorts of reasons — including physical and psychological problems, stress reduction, spiritual growth, and just plain old enjoyment. Most of the bodywork health methods are expressly touch-intensive, with many explicitly acknowledging the therapeutic benefits of touch in and of itself. Because of this emphasis, they're quite attractive for those alienated by the cold, impersonal orientation of modern medicine.

Bodywork offers a wide variety of approaches to choose from. Many of the systems combine movement, exercise, breathing or meditation with hands-on manipulation. The manipulations themselves run the gamut from deep, often painful wrenching of muscle tissue and tendons, to light, feathery touching. We won't be delving into deep manipulations like Rolfing here; while these techniques can be quite therapeutic, they can hardly be described as joyous.

Some of the bodywork systems have fairly long histories, while others are relatively recent arrivals. Despite the variety, almost all of them have one basic aim: to enhance energy flow in the body. Many of them are based upon some notion of body armor, the formulation of psychologist Willhelm Reich. Reich felt that repressed desires and needs become trapped inside different areas of the body. The movements associated with these forbidden drives become frozen in the tissues over many years. This causes tight, shortened muscles which can affect proper posture and movement.

Reich felt that manipulating the body armor zones will force the repressed emotions to the surface. For example, anger may be blocked in the stomach area, so manipulating that spot could release a flood of pent-up frustration. Memories of the childhood beginnings of the repression may also be released. Therefore, working on the body can create emotional catharsis. Non-Reichian approaches also focus on blockages, but don't concern themselves so much with psychological origins. These systems usually derive from Oriental and occult teachings which view sickness as an obstruction of the life force in the body. Whether for psychical, psychological, or other reasons, the life energy in the universe becomes trapped, and the healer's job is to restore the free flow. Reflexology and Shiatsu, as we've seen, employ a similar health model.

The sheer number and complexity of bodywork techniques makes a comprehensive survey impossible here. Rather than cover painful, intrusive therapies — and systems which are inextricably entwined with such non-touch activities as exercise, breathing, etc. — the two techniques discussed here lend themselves nicely to a simplified, do-it-yourself approach, ideal for couples.

Touch For Health Touch For Health (TFH) was this author's introduction to the wide world of bodywork, with several hours of demonstrations and hands-on instruction by Ms. Paula Olecka, a New York City practitioner. She proved to be quite knowledgeable about a number of touch therapies, and provided useful information about some of the other subjects in this book. In Touch For Health, the key word may well be "balance" because the technique focuses on balancing energies throughout the body, which is viewed from a total perspective recognizing how much seemingly unrelated parts can affect each other.

Ms. Olecka proved this rather emphatically in a discussion of a chronic pain in the author's left shoulder. The pain had been increasing lately, and had been interfering with typing ability. Analyzing a daily morning exercise routine, which included leg raises for abdominal muscles, she advised that the leg raises were straining the upper back and neck, thus aggravating the shoulder problem. "Switch to sit-ups," she recommends. That was greeted with some skepticism, because it seemed that leg raises should affect only the lower, not the upper back. But right she was. The shift to sit-ups dramatically reduced the shoulder irritation. After years of practice, a TFH instructor becomes intuitively aware of such not-so-obvious imbalances.

Like other techniques we've looked into, TFH works with the invisible pathways throughout the body called meridians. As with all meridian systems, the basic goal is to stimulate the energy flow in these pathways, thus restoring health and harmony. However, TFH accomplishes this result with different methods, since it relies upon applications from Western disciplines like applied kinesiology and chiropractic. TFH tests for trouble along the meridians by gauging the strength of key muscles associated with each meridian. If a key muscle is weak, TFH techniques will strengthen it. Doing this helps restore the energy flow in the associated meridian. Strengthening weak muscles also helps to balance the musculature throughout the entire body.

The TFH bible is the *Touch For Health* manual, an illustrated, information-packed guide written by Dr. John F. Thie. Thie, who is active in chiropractic and applied kinesiology organizations, established the Touch For Health Foundation in Pasadena, California. The State of California recognizes the Foundation as a Vocational Training School, and accepts TFH coursework as Continuing Education credit for license renewal requirements for registered nurses.

The *Touch For Health* manual outlines a variety of techniques for working with 42 separate muscles. There's no way we can cover anything like that here. Instead, we'll focus on the 14 key muscles linked to the 14 meridians that TFH utilizes. TFH employs four main techniques for strengthening weak muscles. One of them involves deep manipulations of tender spots throughout the body called "neuro-lymphatic points." This can often be quite uncomfortable, so it has been omitted from this book. Another technique entails holding certain acupressure points associated with each meridian. Since this is similar to other acupressure methods already described in great detail, it has likewise been omitted.

The TFH techniques discussed below include "neuro-vascular" holding points (usually found on the head) and meridian tracing. The neuro-vascular holding technique is quite simple. You place the pads of your fingers over the points, and you soon feel a slight pulse. You hold the points until the pulse feels synchronized between the points, then continue to hold the synchronized beat for another 20 seconds, or as long as the condition requires. When you're working with a single point, you just hold it for a while, depending upon the seriousness of the problem. Stimulating the neuro-vascular points seems to improve blood flow to both the muscle being strengthened and to internal organs along its meridian path.

In meridian tracing, you stimulate the energy flow to a muscle by tracing the meridian line which connects with it. Meridians are located on both sides of the body, and you use the flat part of your hand to do the tracing. If the muscles don't strengthen this way, you can try retracing in the *opposite* direction of the meridian line.

The key to Touch For Health is muscle testing. The tests are designed to isolate each muscle being probed. Testing is done by placing the muscle in a certain position (for example, the arm might be held out straight). Then the tester applies a gradual, steady pressure (not a sudden, great force) in a direction opposed by the muscle, and releases the pressure gradually. You only test the first few inches of the muscle's range. If the muscle is strong, it will lock in place in that first few inches. If it's weak, it will give way completely (for example, the upheld arm would swing easily if pressed down).

Naturally, you have to make allowances for the person's natural strength. It may not be easy to detect weaknesses in a strong individual. What counts is *relative* weakness. For example, there should not be more than 15% difference between the right and left sides, no matter which hand is dominant. When testing, look for a lock-in response (strength) or mushyness (weakness). A borderline response should be retested. If the muscle is sound, it should repeatedly retest strong.

If a muscle tests strong, move on to the next one in the cycle described below. If it is weak, do the strengthening exercises and then retest. It should retest strong but, if not, then try the exercises again. As mentioned, an additional strengthening technique is to trace the meridian line in the *opposite* direction than the instructions indicate. If the muscle is still weak, it may indicate a serious problem there or in the associated meridian. Of course, this simplified TFH routine leaves out many other techniques that could be used in problem cases. Consult the manual for complete information.

Since most tests are done with the person lying down, a massage table comes in handy. You can also work on the floor, especially with a thick rug or pad. Don't use a bed, because its softness will interfere with the testing. There are no special preparations apart from this. The entire treatment can be done fully clothed. TFH is intended as a health therapy, not a sensual experience. However, it does have sensual aspects. The holding and tracing are pleasant enough, so there's no reason *not* to approach the treatment in a sensual way. A relaxed setting, soft music and some deep-breathing all act as mood-enchancers. Use your creative imagination.

The routine described below is adapted from the TFH manual. It should be done in the order given, because it follows the natural cyclic flow of energy from meridian to meridian. Obviously, avoid testing any area which is injured or diseased. If testing causes pain, stop immediately and do the strengthening exercises for the painful part. By the time you finish the routine, you'll have charged up weak energy systems throughout the entire body.

The Touch For Health Cycle 1. The first muscle tested is the *supraspinatus*, which helps move the arm away from the body. Shoulder problems are often involved here. Testing: Have you partner stand up, with arm held about 15 degrees away from the body, slightly to the side. The elbow should be straight. You'll be pressing against his forearm to try to push it toward the groin. Holding Points: There are three points here. The first two are above each eye, midway between the eyebrows and hairline. This is the frontal eminence of the brain. The third point is the *anterior* *fontanel*, the baby's soft spot on top of the head. Meridian Line: Trace right on up the center of the body, from the pubic area, under the chin, to just below the lower lip.

Remember, you only do the strengthening exercises (holding and testing) if the muscle tests weak. Also, remember to test *both* sides for each muscle. Most of the meridians run on each side of the body. So if you're testing a left side muscle, you trace the meridian on the left side. You reverse this for a right side muscle. In a few cases (i.e., the supraspinatus), the meridian is located in the *center* of the body, so you use it to treat *both* sides.

2. The second muscle in the cycle is the *teres major* in back of the shoulder. It draws the arm in and keeps it turned out. Weakness here may reflect problems with the upper spine. Testing: The person should lie face down. Have him place a fist against his lower back, elbow bent. Press against the elbow, trying to push it out and down. Holding Points: On each temple at the hairline, slightly to the front and above the ear. Meridian Line: Trace from the tailbone in a straight line down the center of the back, over the top of the head, and down the face to the middle of the upper lip.

3. The third muscle is the *pectoralis major clavicular*, which helps bend and twist the arm at the shoulder. This muscle reflexes to both the stomach and emotional centers of the brain. Digestive problems prompted by stress may be involved here. Testing: The person lies on his back with his arm held straight out, shoulder-level. The palms are out to the side, and the thumb is out towards the foot. Press on the forearm to pull the arm down and away from the body. Holding Points: On the frontal eminence, two points, one over each eye, midway between the eyebrows and hairline. The eminence is often felt as a slight bulge. Meridian Line: Start below the eye. Trace under the eye, up to the forehead, then loop down the jaw and front of the neck, over the collar bone, chest, abdomen, down the outside front of the leg to the end of the second toe. This is the first of the twelve *bilateral* meridians. Remember, trace the *same* side as the muscle you're strengthening is on.

4. The next muscle is the *latissimus dorsi*, which keeps the shoulders down and straightens the back. It reflexes to the pancreas, so weaknesses can reflect such insulin-related problems as diabetes. Testing: The person stands up, arms straight down at the sides. The palm faces away from the body. Press at the forearm to pull the arm out to the side, away from the body. Holding Points: On each side of the head, just above and behind the ears. This is the parietal bone. Meridian Line: Starting at the big toe, trace up inside the leg, over the abdomen and chest. When you reach the shoulder, move down again to the side, just below the armpit.

5. The next muscle is the *subscapularis*, which is behind the shoulder blades. It reflexes to the heart. Testing: The person can sit or stand up. His arm should be out to the side, elbow bent and at shoulder-level. The hand points down toward the feet. Push against the wrist, trying to move the arm forward, the hand towards the head. Holding Points: The baby's soft spot on top of the head, the anterior fontanel. Meridian Line: Trace down the side of the arm, starting at the armpit and ending at the tip of the little finger.

6. The *quadriceps* muscle strengthens the knee and flexes the thigh. It reflexes to parts of the intestines, and can be implicated in digestive disorders. Testing: The

person lies on his back, thigh raised not quite at a right angle, knee slightly bent. Support the heel in one of your hands. With the other, press against the thigh near the knee to try to straighten the leg and push it down. Don't let the leg twist, because this tests a different muscle. Holding Points: The parietal eminence on each side of the head. This is usually a ridge between the ear and top of the head. Meridian Line: Start at the end of the little finger. Trace up the back of the arm, over the back of the shoulders, up the side of the head, and around to the cheek in front.

7. The *peroneus* muscle flexes the side of the foot up and out. Testing: The person should lie on his back. He will turn his toes to the side, with the little toe flexed up toward the head. Secure the foot by holding the heel in one hand. With the other, press against the side of the foot just behind the toes, moving toward the midline and down. Holding Points: The first is the frontal eminence, the same as muscle #3. The second is the glabella, a flat spot on the inside edge of each eye, just above the eyebrow. Meridian Line: There are two moves here. One: trace from the corner of the eye, over the head and down the back to the buttocks. Two: starting at the shoulder, trace down the outer back, over the back of the leg to the end of the little toe.

8. The *psoas* muscle helps to flex the hip. Weakness here can be linked to low back pain, kidney ailments and foot problems. Testing: Lying on his back, the person raises his leg about 45 degrees, slightly to the inside, foot pointing out. Press against the inside of the ankle, trying to move the leg out and down. Holding Points: The prominent lump in the back of the head near the base of the skull. Meridian Line: Begin at the ball of the foot. Trace up the inside of the leg, up the center of the body to the middle of the chest.

9. The *gluteus madius* pulls the thigh out and rotates the leg. It reflexes to the sex organs, and can be implicated in impotency, prostate and menstrual problems. Testing: The person lies face up, legs straight out to the side. Secure one ankle with one of your hands, as you push on the opposite ankle with your other hand, trying to move it toward the midline. Make sure the hips don't rotate in this position. Holding Points: Same as muscle #6. Meridian Line: Trace from one nipple down the middle of the inside of the arm to the end of the middle finger.

10. The *teres minor* muscle rotates the arm and forearm. Many conditions can be connected with a weakness here, including wrist and elbow problems, digestive disorders and thyroid imbalances. Testing: The arm is held near the side, with the elbow bent at a 90-degree angle, thumb toward the shoulder. The forearm is turned out as far as possible. Secure the arm at the elbow with one hand, while you press with the other against the back of the wrist, trying to push the forearm across the chest. Holding Points: One: just above and to the front of the ear, on the temple at the hairline. Two: in the depression just above the breast bone. Hold points one and two *simultaneously* with each hand. Meridian Line: Beginning at the tip of the ring finger, trace up the back of the arm, up the shoulder and side of the neck, around the ear to the eyebrow.

11. The *anterior deltoid* helps flex the shoulder when the elbow is bent. It reflexes to the gall bladder, so weaknesses in it can be connected with overindulgence in fatty foods. Testing: The person lies face up, arm straight out at a 45-degree angle above the leg. The palm is down. Press against the forearm to try to push the arm

down. Holding Points: Same as muscle #5. Meridian Line: Starting at the corner of one eye, looping around the temple, behind the ear, up again to the forehead, then down the back of the head and shoulder, side of the chest, outside of the leg to the tip of the fourth toe.

12. The *pectoralis major sternal* moves the arm in. Weaknesses here can be reflected in liver and eye trouble. Testing: The person lies face up, arm held straight forward and slightly to the side, shoulder-level. Palm should be out, thumb towards the feet. Push on the forearm to move it towards the head and out. Holding Points: Two spots along the hairline about one and a half inches to each side of center (more or less above each eye). Meridian Line: From the big toe, trace up the front, inside of the leg, over the stomach to the lower portion of the chest near the side.

13. The *anterior serratus* muscle draws the shoulder blade forward and raises the ribs. Weaknesses here can manifest in shoulder blade and lung problems. Testing: Seated or standing, the person should hold his arm straight out, shoulder-level, 45 degrees to one side. The thumb should be up. Use one of your hands to hold the tip of the shoulder blade so it can't slide down. With the other hand, press against the forearm to move it down towards the floor. Holding Points: Same as muscle #5. Meridian Line: Starting at the chest, trace down the outside front of the arm to the thumb.

14. The last muscle tested is the *fascia lata,* which helps bend the thigh and move it sideways. Weaknesses here can be connected to intestinal problems and chest soreness. Testing: Lying on his back, the person should raise his leg up to a 45-degree angle, a little to the side. The foot should be turned in. Press against the outside of the ankle, trying to move the leg down and in. Holding Points: Same as muscle #6. Meridian Line: From the tip of the index finger, trace up the outside back of the arm, over the shoulder and jaw, up to the nose.

That's the full TFH cycle. We've covered the 14 major meridians in the body. After boosting all the weak muscle reflexes, the body should feel rejuvenated and raring to go. Regularly receiving TFH treatments may promote a healthier, bouncier life, a sort of "apple a day keeps the doctor away" effect. Trade the treatments with your touch-partner, and check out the results for yourself.

One final TFH technique is worth mentioning. You can do this one all by yourself. It works well if you feel angry, frustrated, or tense. Place the first two fingers of each hand on each side of the frontal eminence points on your head (between eyebrows and hairline). Apply constant light pressure with both sets of fingers simultaneously. While holding the points, mentally review the problem which is causing your turmoil. Let go with your hands, then repeat the entire process—holding, mentally reviewing, releasing—twice more.

By that time, much of the negative feelings should clear up. If you're trying this with someone else, you can review the difficulty verbally while the other person holds the points. This can be quite soothing. The technique is also useful for nightmares, fears, and other emotional or mental strains.

Polarity Therapy Polarity is a marvelously simple healing system which just about anyone can master. It's also a fine bridge between very physical techniques, like acupressure and TFH, and the subtler methods of psychic healing dis-

Frontal eminence points

cussed in the next chapter. Polarity manipulations can be considered a cross between psychic healing and more traditional types of massage.

Polarity therapy was pioneered by Dr. Randolph Stone, an Austrian who immigrated to America in the early 20th Century. After studying several ancient healing systems, he developed and refined his own theory over the course of sixty years.

Dr. Tom Florian, a Virginia Beach chiropractor and polarity practitioner, provided a sample of his copy-righted paper, "An Introduction to Polarity Therapy," and also suggested *Your Healing Hands: The Polarity Experience*, a popular account by Richard Gordon. The techniques presented here are a composite of Florian's and Gordon's work.

Like many of the systems already discussed, polarity considers the body to be sustained by an electro-magnetic type "life force." This life force permeates the body through invisible channels. When the circulation is blocked because of excess stress or other conditions, disease results. The polarity therapist's job is to remove energy blocks by getting the channels flowing freely again.

Polarity principles divide the body into poles much like that of a magnet: the top of the body and right side are *positively* charged, while the feet and left side are *negatively* charged. These charges are based upon the positive and negative energies of the life force.

The polarity practitioner uses these charges to balance the body's life energies. He heals by transferring positive and negative stimuli from his own body to his subject's. When the life energies are realigned and balanced, the nerves become quiet and soothed. Tense muscles let go, which sometimes allows pinched bones to slip back into their proper places. The main benefit of a polarity session is a deep, thorough relaxation. Actually, the benefits are boosted the more someone is "out of sync" mentally or physically. The really upset or ill individual is most in need of balancing. However, even a healthy, stable person can be made to feel peaceful and relaxed.

Essentially, the polarity technique utilizes the therapist's hands as a sort of battery. Since the right and left hands have opposite charges (positive and negative, respectively), they can be used as opposite poles of an energy circuit. Placing them on certain sensitive spots on the body acts as a "jumper." The positive/right and negative/left form an electro-magnetic type connection between any parts they link. The life energies freely flow between the healer's hands, thus rejuvenating and refreshing the receiver.

To get some idea of the magnetic quality of the hands, hold your palms a few inches apart for awhile. Now move them closer, then further apart several times. Eventually you'll feel some sensation like tingling, prickling, temperature change or attraction. The sensation can be strong or subtle. According to polarity therapists, what you're feeling is the life energy being drawn between the opposite poles of your hands. You create a similar current when you use your hands to give a polarity treatment.

Preparations The person giving the treatment should generally be in a good frame of mind. Since he will be transferring energy, it's important that he not be harried, hurried, or negative. A relaxed outlook should be the aim of *both* parties. Before starting, it's not a bad idea for giver *and* receiver to do a few minutes of deep breathing and meditation, releasing any distracting, disturbing thoughts. Deep, even breathing by both parties should be continued during the session itself, since it enhances energy flow.

Naturally, the setting should also contribute to a mellow state of mind. Quiet, comfortable surroundings are best. Keep distractions and potential interruptions to a minimum. Subdued lighting, soft music and incense are appropriate. The polarity session is a form of massage, so you want to create the kind of relaxed mood you would have for massage. Clothing worn by both parties should be loose and comfortable. Nudity is often helpful, since it eliminates obstacles to energy transfer. At a minimum, the receiver should remove shoes and socks, jewelry and any other metal objects — belt buckles, keys, etc. Metal affects electromagnetic flow.

A massage table is probably the best place in which to do polarity work. It allows the giver to move around freely (see the massage section for tips on makeshift massage tables). The next best bet is a comfortably-padded floor. Don't bother with beds — they're too soft to maneuver on. Also, wait at least an hour after eating a main meal before giving or receiving a session.

Basic Moves Ten polarity positions have been selected for coverage here, from the dozens available. Omitted are movements that involve uncomfortably intrusive pressure, as well as those which are basically manipulations of joints. Rather, we shall discuss the ones which best facilitate the flow of energy. Follow the sequence given here, since it is designed to produce a complete energy exchange. Follow the instructions closely; hand position is the key to polarity. Generally, the right/positive hand is placed on a negatively-charged body spot, while the left/negative hand is liked to a positively-charged place.

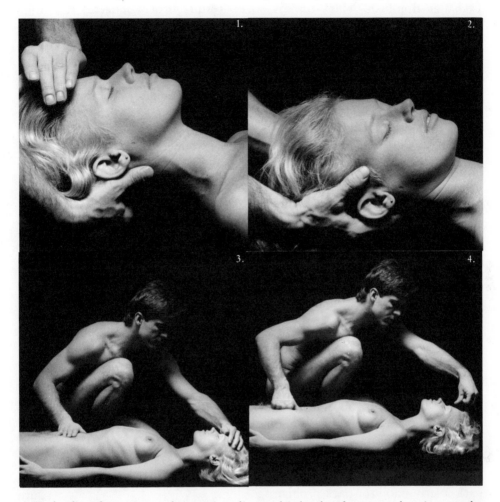

1. In the first few moves, the receiver lies on his back. If you are the giver, make sure your hands are warmed up before beginning. Rub them together briskly. Now place your right hand lightly on the back of your partner's neck. Your left hand should be about half an inch over (not on) his forehead. Ask him to take a number of deep breaths. You should feel a concentration of energy in your left hand. Leave your hands there until the sensation of energy transfer subsides. This position effectively balances the energy in the skull, and is especialy good for headaches.

2. Cradle your friend's head underneath the palms of your hands, fingers toward the base of the skull, thumbs above the ears. You should feel pockets of tension around the neck. *Use light pressure here.* Hold the position until the tension lessens, and the energy exchange diminishes. This is a soothing, relaxing position, also good for headaches. You can vary this movement by applying deep pressure with your fingers on tension pockets at the base of the skull and neck. Alternate between pressing and letting go until the tension lessens.

3. Position yourself on your partner's right side. Put your left hand on the forehead, and your right on the stomach. Rock the person back and forth with your right hand. The rocking should be rhythmic and gentle. Do it for a few minutes, then stop and leave your hands in place until you feel the energy exchange diminish. This "tummy-rock" results in a wonderful relaxation which works well before bedtime.

4. Stay on your partner's right side. Make fists with both hands, thumbs pointing down. Lightly rest the right thumbs slightly below the navel. Put the left thumb less than an inch *above* (not touching) the eyebrows, toward the center of the forehead. Hold the posture for a few minutes. You should feel a tingling energy transfer on your left thumb.

5. On the right side of your partner, hold his left foot in your right hand, and his right hand in your left. Hold until the energy flow slows down. Then move to the left side, holding his right foot in your left hand, and his left hand in your right.

6. With both hands, brush down one of your friend's legs, from above the knee to the toes. When you reach the toes, shake your hands as if you were throwing off water. Repeat several times, then do the other leg. This movement helps circulate stagnate energy, and the brush-off action keeps you from returning it again.

7. Polarity therapists usually do foot massage as part of their routine. You can use thumbs, fingers or knuckles to rub the entire sole. Do the heel, sides, and top of each foot. Massage the toes, then pull on each one, trying to make it pop. Don't worry if any don't pop, the main thing is to give a good tug. Do both feet.

8. Your partner will now turn over so that he is lying on his stomach. Position yourself by his left side. Put your left hand on the bottom of his neck, and your right on the base of his spine. Rock gently with your right for a couple of minutes, then keep your hands in place for awhile. Now move your right hand up to the lower back, and hold unitl you feel your friend relaxing. Continue to move your right hand up the spine to the upper back. Move an inch or so at a time, and hold for a while between each position. This set of movements is particularly good for back problems.

9. Place the middle finger of your right hand under the tailbone. Your finger will rest near the outer edge of the anus. Your left hand will be placed on the upper part of the spine. Hold this position for several minutes. This movement sends the life energy pulsing up the spine, causing some rejuvenation.

10. Finish off the session with back and front "brushes." Have your partner sit up straight. Get behind him, and place your right hand on his right shoulder, your left hand on his left shoulder. Have your right hand over the top of the back toward his left side, while your left hand moves in the opposite direction. Your hands will cross in the center of the back. Continue the movement down the sides, with your right hand now on his left side, your left hand on his right side. When you reach the bottom of the back, move both hands toward the base of the spine, where they will cross again. Shake your hands briskly after each complete movement. Repeat the procedure about ten times, lightening the pressure each time, until you are not touching at all.

Now move to the front side of your seated friend. Pleace each palm on one side of his head. Brush down the entire body, left hand over his right side, right

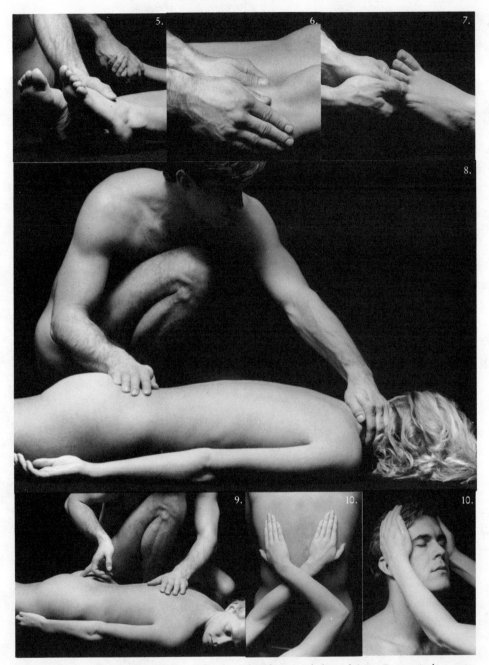

hand over his left side. Shake your hands after each brushing. Repeat about ten times.

That's the end of the short-form polarity session. Even with the lesser number of movements, the effect should still be deeply relaxing and refreshing. The receiver may want to rest after the treatment. The giver should wash his hands in cold water immediately after. This helps "ground" any negative or sick energies picked up from the receiver. For the same reason, it is advisable that the giver shake his hands after each movement during the session.

Polarity is one of the simplest full-body relaxation aids available. Nor does it take much out of the receiver physically. This makes it easy for a couple to trade polarity treatments in the same session. Try it.

CHAPTER SIXTEEN: PSYCHIC HEALING

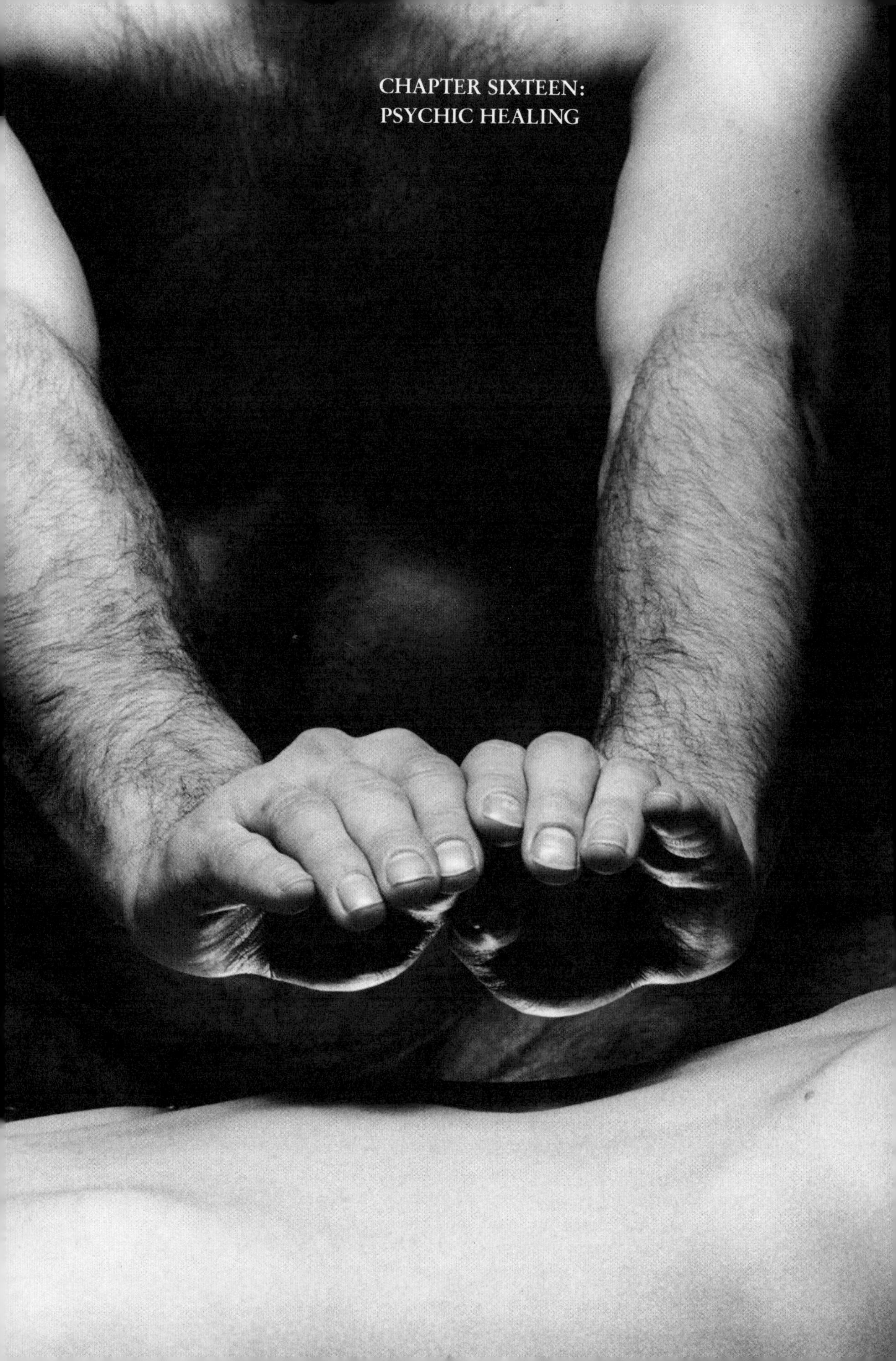

You may wonder what psychic healing has to do with touch. After all, isn't psychic healing purely in the realm of the mind? Isn't it inextricably wrapped up with religion, faith and spiritual beliefs?

Yes, indeed, some types of psychic healing *are* purely spiritual. These are the miracle cures which defy explanation: the type of healing in which the blind suddenly see and the lame suddenly walk. Those cures are the province of 'saints. What we are dealing with here is much more modest, and has been practiced by common folk from the dawn of cilivization. The technique has had a number of names, including "laying on of hands," "mesmeric passes," and "animal magnetism."

And it is definitely a touch technique. Commonly, the hands make direct contat with the person to be healed. However, this is not absolutely necessary, for some healers sweep close to but don't actually touch the body. Even then, the receiver feels an energy pulsation affecting his body. This is one of the intriguing aspects of psychic healing—it provides "non-physical" touch. The healer can not only make contact without direct touch, but the receiver can "feel" the non-contact movements as well. In psychic healing, you *don't* end at your skin.

Of course, none of this has been proved with the precision of a Ph.D. dissertation. In fact, the conventional medical establishment would likely scoff at psychic healing, just as they do *any* uncommon health system. Actually, many tests have been run in respectable labs demonstrating the effects of psychic energy. Electrophotographs made of healers' hands, as well as of their subjects, have shown definite energy shifts after healing. The evidence is not overwhelming, but it is deserving of consideration.

It's worth repeating that the psychic healing we're talking about here is *not* a miracle cure. What is said to occur is an energy transfer between giver and receiver. Like all other health systems discussed in Part IV, psychic healing views health as the free flow of universal life energy in the body. This energy—ki, prana, orgone, or whatever term you prefer—is constantly replenished through breathing, eating, drinking, and other nourishment. However, poor diet, psychological problems or other negative conditions can block the energy flow, causing disharmony and disease.

The psychic "healer" actually heals nothing by himself. His role is to act as a *channel* to help clear the blocked energy passages. He provides a "boost" or "jump" to get the patient's own inner processes balanced again. The healing itself is done by the patient; the healer is the catalyst who gets it going.

The actual energy exchange usually occurs through the fingertips of the healer. He will often feel a pulsing sensation of energy departing from there, and the patient generally perceives this as a tingling or temperature change on his own skin. Labs in both the U.S. and U.S.S.R. have recorded emanations from the fingers of famous psychics. The researchers employ Kirlian photography, which uses an electrical charge to reveal energy pulsations around objects. Dr. Thelma Moss of UCLA has taken Kirlian photos of the fingertips of psychics Uri Geller and Olga Worrall while they demonstrated their techniques. The shots clearly show energy shooting from the fingers of both. Other Kirlian photos of healers show that the fingertip emanations lessen after the healing is completed. In contrast, shots of their patients show *enhanced* energy after a healing session. Indeed, it would seem that some sort of transfer *is* occurring.

Many observers believe that the Kirlian photos demonstrate the existence of the ancient notion of the human aura. Occult teachings say that each person's body is surrounded by a corona of energy. Psychics are supposed to be able to clearly see this aura, and their descriptions often match what Kirlian photographs pick up. Some books describe a technique by which *anyone* can get a glimpse of the aura: have someone stand in front of a white or neutral-colored wall in a dimly-lit room. What will you see? It could be a faint hue surrounding the figure, or colors shooting out from it or smoke-like emanations coming out of the fingers.

The aura is often described as the source of the mysterious healing life-energy (ki, prana, etc.). "Persons of psychic sight see plainly the spark-life particles of Prana being shaken from the finger-tips of persons giving so-called 'magnetic treatments' or making mesmeric passes." (Yogi Ramacharaka, *Fourteen Lessons in Yogi Philosophy*).

In fact, some respected thinkers consider the aura to be not simply an extension of the physical body, but a *separate* energy core which co-exists with the body. It is believed to be slightly larger than the human form, so it extends beyond, leading to the corona-like effect. While occultists call this an "astral" or "etheric" body, Soviet researchers use the more scientific-sounding "biological plasma body." The Soviets claim that all living creatures have this twin energy body, and that it may be the explanatory lnk for paranormal occurrences like ESP. It is noteworthy that Kirlian photos of dying plants and animals show the corona energy shipping away, and then disappearing entirely after death. Can this be evidence of the "spirit" of "soul" departing? It's certainly an intriguing subject to speculate on.

Nonetheless, you needn't believe in auras, astral or bioplasmic bodies to do psychic healing. Faith in *anything* is by no means necessary for this purpose. Of course, you won't get that impression from famous faith-healers, who attribute the strength of their cures to the patient's belief. Faith may be important in miracle cures or self-hypnosis, but not in simple energy transfer. The least esoteric explanation for psychic healing is that it involves an exchange between the electro-magnetic fields of people. Several researchers in the U.S. have used sensitive voltmeters to record such electro-magnetic fields around all living things.

In *Magnetism And Its Effects On The Living System*, researchers Davis and Rawls took electrical measurements of the entire human body, from head to toe. They found that the body is polarized much like a magnet. Just as polarity therapists would maintain, Davis and Rawls discovered that the right palm has a positive charge, the left, negative. They also learned that the back of the body is positive, the front negative.

In a subsequent book, *Rainbow In Your Hands*, Davis and Rawls set out to do a scientific study of the effects of "laying-on of hands." They found that using right palm (positive) energy strengthens, while left palm (negative) energy reduces. (The back of the hands, by the way, have the opposite charges: right/negative and left/positive). In healing, this means that the right palm increases and boosts energy, while the left soothes, sedates and relaxes. The left is the pain reliever, while the right can actually *increase* pain by stimulating the nerves.

Used simultaneously, the two palms complement each other, the right strengthening and the left sedating. If you put your palms together, you produce a closed energy circuit, which may explain the prevalence of hand-clasping in prayer. Davis and Rawls recommend rubbing palms together to enhance the energy flow before beginning a healing session. They also suggest placing one hand over an injured spot, and the other on the opposite side of it. This creates the closed crcuit of positive-negative healing energies.

Psychic Touch Techniques In the late 18th Century, an Austrian physician, Anton Mesmer, developed his "animal magnetism" healing techique. The term "mesmeric passes" is also used to describe it. While chastised by the medical establishment of his day, he had no lack of satisfied patients. Mesmer's "Maxims on Animal Magnetism" have been translated and reproduced in Jerome Eden's interesting book, *Animal Magnetism and the Life Energy*.

Mesmer's treatment starts with the therapist directly facing the patient. He then puts his hands on the patient's shoulders, and traces down the arm to the fingertips. He holds the patient's thumbs for a moment, then repeats the entire movement several times. The tracing is done with the thumb and index finger, or the palm, or one finger reinforced by another, or all five finers. Next, the hands are placed on the patient's head, and a second series of sweeps pass all the way down to the feet.

After these prelminary moves, the therapist focuses on sick or injured parts. Mesmer also utilized polarity principles. He suggests touching an injured part with the right hand, while placing the left hand on the opposite side. Mesmer's treatments included a prolonged, concentrated touching of affected areas, usually spread over a number of sessions. Initially, the bombardment of magnetism will *increase* the symptoms or pain, as the blocked energy in the disturbed spot begins to churn. Through repeated attempts, the dam bursts, and the energy obstruction is eliminated. Mesmer called this a "crisis," and said it can often be accomplished by violent reactions by the patient. After the crisis, energy circulation is restored, and the patient is cured.

Volume 1 of Yogi Ramacharaka's *Fourteen Lessons in Yogi Philosophy* (written around the turn of the century) outlines these ancient techniques:
1. Stand in front of your seated patient. Raise your arms level with his head, palms toward him, fingers outstretched. Sweep your arms slowly toward his feet. When you reach them, turn your palms toward yourself, and reverse the sweeping up to his head. Repeat several times, then concentrate the same up and down movements over any disturbed areas. After treating them, flick your fingers as though you were tossing off water, to avoid picking up any sick energy.
2. Chronic or long-standing conditions can be "loosened up" with "sideways passes" before the affected part. Stand facing the patient, with your hands together, palms touching. Then swing your arms out sideways several times. Follow this with the downward sweeps already discussed.
3. Headache treatment: stand behind your seated subject. Spread your fingers

open and down, then pass them over the top of his head in double circles, one circle with each hand. Don't actually touch his head. After a short while, you should feel energy passing from your fingers to his head. Another soothing move is to hold your hands on the patient's head, over the temples.

4. Pain treament: put one palm in front of the afflicted part, a few inches away. Hold it there for several seconds, then slowly rotate it round and round over the seat of the pain. To stimulate circulation in a painful spot, point a forefinger toward it, a few inches away. Keeping the finger pointing steadily, move your hand around as though you were boring a hole with the finger-point.

5. Breath technique: some healers breathe right on the affected area. Often, a cotton cloth is placed between the flesh and the healer, with the breath warming up the cloth. Cotton has been claimed to be an especially good conductor of healing energies.

6. Ramacharaka feels that much of the value of massage comes from energy transfer. The effect can be enhanced when the person giving the massage consciously directs the energy flow from his fingertips. A closer look at psychic massage is included below.

Dolores Krieger, who is both a Ph.D. and Registered Nurse, teaches a unique course as part of the graduate program in nursing at New York University. The course outlines the system she calls "Therapeutic Touch," a modern derivation of laying-on of hands. Dr. Krieger has studied with both scientists and psychics, and is by no means a wild-eyed occultist. Her technique has been analyzed in laboratory experiments, where it was found to produce profound relaxation in her subjects. Dr. Krieger says that several hospitals throughout the country have Therapeutic Touch teams which combine traditional methods wih the new technique. These teams use Therapeutic Touch for various purposes, including alleviating patients' pain and anxiety, and speeding up the healing process in fractures.

In her book, *Therapeutic Touch*, Dr. Krieger lists the key uses of the technique as pain relief and deep relaxation. She feels the benefits stem from sparking the natural inner healing processes we all posses. The energy transferred from the therapist's hands enhances and balances the patient's energy flow. In Dr. Krieger's view, everyone has an innate ability to heal this way. She has successfully taught the technique to scores of nursing students, and feels that almost anyone can pick it up with practice. What follows is a condensation of the Therapeutic Touch (TT) technique described in her book.

Dr. Krieger suggests getting a feeling for the human energy field by practicing with your palms. Have your palms face each other a few inches apart. Maintain that position for a while, then move them a few more inches apart, hold again, then move another couple of inches apart. Then reverse the process, moving and holding them a few inches closer each time. You should begin to feel a certain subtle pressure impinging on your palms. This is the energy field. You'll be looking for a similar sensation when you work on someone else's field.

Before starting a TT session, Dr. Krieger emphasizes that you center yourself. Centering is an important concept in all psychic healing, and should proceed any of the techniques in this chapter. The idea behind centering is that the person doing healing should be in harmony with himself. A healthy, non-hassled state of mind is needed to help someone else. Otherwise, you cannot seriously expect to balance someone else. Basically, centering involves attaining a relaxed, peaceful, internal state. Do this by closing your eyes while sitting in a comfortable position. Breathe deeply and evenly, and allow each muscle in your body to let go and relax.

Focus on each body part from head to toe, and allow each one to release any tension within. Continue to breathe deeply and evenly. If you know how to meditate, this is a good time to practice it. Eventually, you'll feel a center of balance and gravity inside you. In most people, this is around the solar plexus or navel. Let your mind settle deep within, as your body comfortably attains equilibrium around its center. Feel as if your consciousness is flowing out of the center. Try to maintain this calm, serene balance throughout he entire session. Dr. Krieger calls this "effortless effort."

The next step after centering is to assess your friend's energy field. Stand in front of your seated subject. Stretch your hands out toward his head, two to three inches from his skin. Move your hands slowly, carefully down his face, looking for differences in sensation between his two sides. A trouble sign is when you feel a change in temperature in your palm over some spot. Take about seven to ten seconds to scan from the top of the head to the chin.

Continue this scanning procedure all the way down the front of the body. Once again, search for any potential trouble spots. Any change in temperature, pressure, or other unusual sensation in your palms will alert you to them. Eventually, you get a feel for what the normal field is like, which makes it easier to pick out abnormalities. When you finish the front, repeat the scanning all the way down the back part of the body, starting at the top of the head. Once again, make mental notes of any disturbed parts. When you complete the entire body, go back and recheck any areas you had doubts about.

The next step starts the actual healing process. Return to the places where you felt unusual sensations. These spots often feel "statical" compared to the rest of the person's field. This static condition indicates congestion of the life energy. To remedy this, you "unruffle the field." Place your palms over the affected area, then move them from the body in a sweeping motion. In most cases, you'll gesture downward, following the lines of the limb nearest the troubled spot. You can also sweep perpendicular to the body.

Dr. Krieger says that the sweeping "feels as though I were actually pushing a pressure front" (*Therapeutic Touch*). Continue the sweeps, and eventually you'll find that the static in the field loosens up — you have "unruf-fled" it. What you've been doing is forcing blocked energy to flow again. This boosts the healing process, and is particularly helpful for pain relief. After finishing the unruffling movements, Dr. Krieger suggests shaking or washing your hands, to avoid contamination by negative energies.

The next step is to balance your partner's energy field. Focus your attention on areas where you felt peculiarities in the energy field. The goal now is to moderate the peculiarities so they're once again in balance with the whole. For example, if you experienced a cold sensation, you want to warm it up. If you felt tingling, you want to soften the vibrations. Whatever the nature of the imbalance, the idea is to tone it down.

You accomplish this by directing energies from your own body to the affected areas. One way to do this is to create a force field between your palms in the manner mentioned above. Then place one palm over the disturbed part, with the other palm on the opposite side of the body. Consciously direct the flow of energy into your friend's body while you mentally visualize the disturbance moderating. A second technique for balancing involves transferring energy between different part of your friend's body. You use your hands as a connecting rod, in much the same way as done in polarity therapy. For example, let's say you detect a disturbance in your partner's knee. Put your right hand on the knee and your left hand on his foot (on the same leg). That way, you create an energy circuit running down the leg. Your left hand/negative energy will draw your right hand/positive energy to it.

Generally, the session ceases when you no longer detect any disturbances in your friend's field. Both sides will feel the same, and you will pick up no special sensations of tingling, temperature, etc. When finished, wash your hands off in cold water.

Naturally, it takes a good deal of experience to interpret these subtle cues. Dr. Krieger suggests regular practice on your friends and relatives, or even animals and plants. When it comes to psychic healing, practice really makes perfect. The powers seem to build with use. The next healing technique may be the most interesting of all, since it combines psychic energy with massage. Roberta DeLong Miller, who formerly headed Esalen's massage department, has developed a system she calls "psychic massage." She outlines her method in a book, appropriately titled, *Psychic Massage*. The following paragraphs present a shortened version of it.

The preparations for psychic massage are much the same as for Therapeutic Touch. The person giving the treatment should center himself first. Then, he should get a feeling for his energy by creating a force field between his palms. Next, he should assess his partner's field, checking it from head to toe.

Now the actual massage movements begin. Your friend should lie on his back, preferably on a massage table. You will be using massage oil, as in a regular massage (Ms. Miller recommends Sesame oil). Rub oil over your hands, then stand behind your partner's head. Close your eyes and feel the vibrations coming from him. A circular flow should start between your connected energy fields.

When you feel this energy flow, slowly move your hands to his chest. Allow them to sink in without pressing, until you sense the energy exchange strengthening. Now massage the entire chest area — ribcage, collar and breastbone, pectoral muscles. Don't be concerned with specific massage moves — do whatever feels natural. You can choose from the many movements mentioned in the massage section above. After finishing the chest, rest your hands on the heart center, which is slightly to the right of the heart. Breathe deeply, and direct your energies into the center. This connects with your partner's emotional being.

Put more oil on your hands, then continue the massage strokes on the shoulders, neck, under the upper back and back of the head. Repeat all moves several times. After thoroughly working these areas, slip your hands under the person and work his shoulder blades and the upper part of his spine. Loosen up the tissues around each vertebra there. Your object is to open up a channel for psychic energy in the spine.

Test this by placing your fingers over the vertebra which sticks out at the base of the neck. You're doing the right job if you feel a strong current flowing into your hands. If you don't feel it, continue loosening the spine until you do. Then move your hands up to the neck vertebrae, and give special attention to any tension-knots you find. Massage the rest of the neck, then work the scalp with your fingertips. Put one palm on top of his head to feel for the energy flow. Then cup your hands under the skull and gently stretch the neck, trying to loosen up the vertebrae.

Next, massage the face, paying particular attention to all tension-spots. Finish up by placing your fingers over the "third eye" (located between and slightly above the eyebrows). Feel for the energy flow there.

Every now and then, take stock of your own psychological state. Make sure you stay centered with even, regular breathing. Then put some more oil on your hands and move to your partner's right side. Extend his right arm out, and thoroughly massage it from shoulder to hand. Use long, light strokes, as well as localized kneading. When finished, hold his hand in yours, and experience the energy exchange between the two of you. Then repeat the process on his left arm. Next, begin clockwise, circular stroking on the abdomen. Ms. Mills recommends making a full circle with the left hand, and a half-circle with the right. Travel from side to side across the abdomen in this fashion. Move up to the diaphragm area, and down to the lower belly. Then work the muscles on both sides.

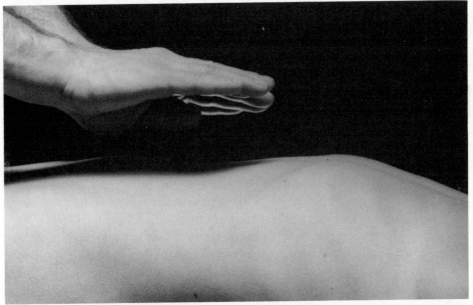

Finish by resting both palms on the belly center, an inch and a half below the navel. Feel for the energy flow here. This is the seat of power and creativity, so a connection here can lead to a potent exchange between you. Ms. Miller puts it this way in *Psychic Massage:* "As your beam of Energy comes into that place, there is a synergistic gain of presence. No longer are you two bodies, but one overwhelming unity."

Next, massage the legs and feet, starting with long, downward strokes from thigh to foot. Then work your way with localized kneading from the top of the thigh on down. Do both legs. After that, hold both feet. Experience the energy from your hands surging throughout his entire body. The vibrating force from all his body parts should surge back to your hands. Let your friend savor this experience for awhile, then have him turn over on his stomach.

Center yourself and try to sense his energy field. Then slowly move your hands through the field to his back. Gently spread oil over his back and legs. Now position yourself beneath his feet so you can work the sole and heel. Next, massage up his leg, over the calf and thigh, using stroking and kneading movements. Move up to the buttocks, working them with sweeping, circular motions. Finish the other leg in the same way. Complete the leg work with long sweeps starting at the heel and extending all the way over the body, then down the arms. You want to get the energy surging from *bottom* to *top* when you work on the back of the body.

Next, work the sides and back from bottom to top with long, circular strokes like the ones you used for the abdomen. Then massage the muscles on both sides of the spine. Now do the area around, under and on top of the shoulder blades. Then use your palms to knead all of the muscles on the upper back. The following move is the most important part of the entire session. Starting at the tailbone, massage the hollow between each vertebra, loosening up the connective tissue. Don't press on the bone itself. Work one side of the spine, then the other. Pay special attention to any vertebrae which seem out of phase, or show an obstructed energy flow. Send energy where it's weak, and soothe over energetic spots. Your aim is to build a smooth energy connection all the way up the spine.

This completes the massage, although you can add your own final touches, like drawing figure eights on the back. Finish up by re-focusing your energy, then position yourself by your friend's head. Bend over and place all your fingers on the base of the spine. Stay there until you connect with his energy, then slowly move up the spine, keeping the connection. Steadily draw his energy up with your hands, taking at least two minutes to completely cover the spine.

"The person will feel you moving his entire being, from his depths to his heights. His body begins to vibrate under your hands, like a sympathetic string harmonizing with you. The experience is overpowering," Ms. Miller says in *Psychic Massage*.

When you reach the head with your fingers, hold your hands there a while. Experience the connection with your friend's entire body. Now draw your hands back an inch from his skin. Both of you should still feel the energy exchange. Move back slowly, inch by inch, until you're more than six inches away. You will then be covering his outer aura. Then remove your hands completely. That is the end of the session.

This shortened version of Ms. Miller's method cannot go into the many extra moves she outlines, nor can it fully capture the spiritual flavor she imparts. Any reader really interested in psychic massage should examine her intriguing book. She also includes many other healing techniques, as well as a unique "psychic view" of human psychology and hang-ups. A thoroughly fascinating perspective.

I'd like to close this chapter by describing my own personal experiences with psychic healing. In the summer of 1980, I attended a half-dozen sessions on the subject taught by Ron Mangravite, a psychic healer who also does serious research as Director of the Jersey Society of Parapsychology. I initially approached his demonstrations with a skeptical "show me" attitude, a product of my background in social science and law. Believe me, he showed me a lot more than I was prepared for.

Mangravite's method for boosting psychic energy was to have the group form a circle, with hands linked. He would then direct his energy around the circle, which many members felt as a tingling inside. The energy seemed to be amplified as it passed from person to person, and the room temperature rapidly rose each time. I experienced a throbbing pulsation in my hands. At one point, he had everyone turn in one direction, and he instructed us to raise our palms and send energy into the back of the person in front. Although we weren't even touching, I soon felt a heat sensation on my back. The guy in front of me turned around and said, "ease up. You're burning the hell out of my back."

Mangravite showed us how to test people's energy field, which was surprisingly simple. As we passed our palms a few inches away from someone's skin, we could often pick up static, disturbed vibrations over distressed spots (we weren't told where they were before testing). He also demonstrated the differences between left and right hand energies. He could actually attract or repel someone's body by using positive or negative energy in the palms!

Mangravite did some intense one-on-one healing work with volunteers in the group. In many cases, when he passed psychic energy into some troubled area, the person would be thrown into involuntary convulsions. One volunteer vomited, another shuddered uncontrollably and wept. These were gripping experiences, and they seemed much like Mesmer's descriptions of the "crisis" which precedes all psychic cures.

But for me, the spaciest scene occurred when he decided to show us his aura. He backed up against a blank wall in dim lighting, then began meditating with his legs crossed on the floor. In a short while, I began seeing multicolored energy pulsing around him. It looked much like a fuzzy, constantly shifting outline of his body. He then told us to watch how the aura changed with his state of mind. Spiritual thoughts produced a calm, smoother outline, while insane thoughts resulted in a jagged, disjointed cast. But the most startling aura effect occurred when he said that he would project his astral body outside his physical self. He aura began to rise up, and several feet above his head appeared a dull glow.

But even *that* wasn't the most startling thing I saw. In another demonstration, he set up an energy field between his palms. I began to spy a pulsing force flowing between them, and this crystallized into a thin beam of amber-colored light. The beam stretched like a rubber band as he moved his hands apart, and collapsed like tubing as he brought them together. I saw the beam as vividly as I've seen anything in my life. Hanging on to my last threads of skepticism, I squinted my eyes, widened them, closed one eye — anything to change my visual perspective. But no matter what I did, the beam remained clear. And I certainly wasn't the only member of the group to view both the beam and aura.

The sessions with Mangravite provided some startling revelations, definitely shaking up one's world view. I'd read plenty about psychic touch before then, but seeing and feeling it in the flesh made me a lot more sure about its reality.

Carefully check out the exercises in this chapter, and I'm certain you'll experience the effects too.

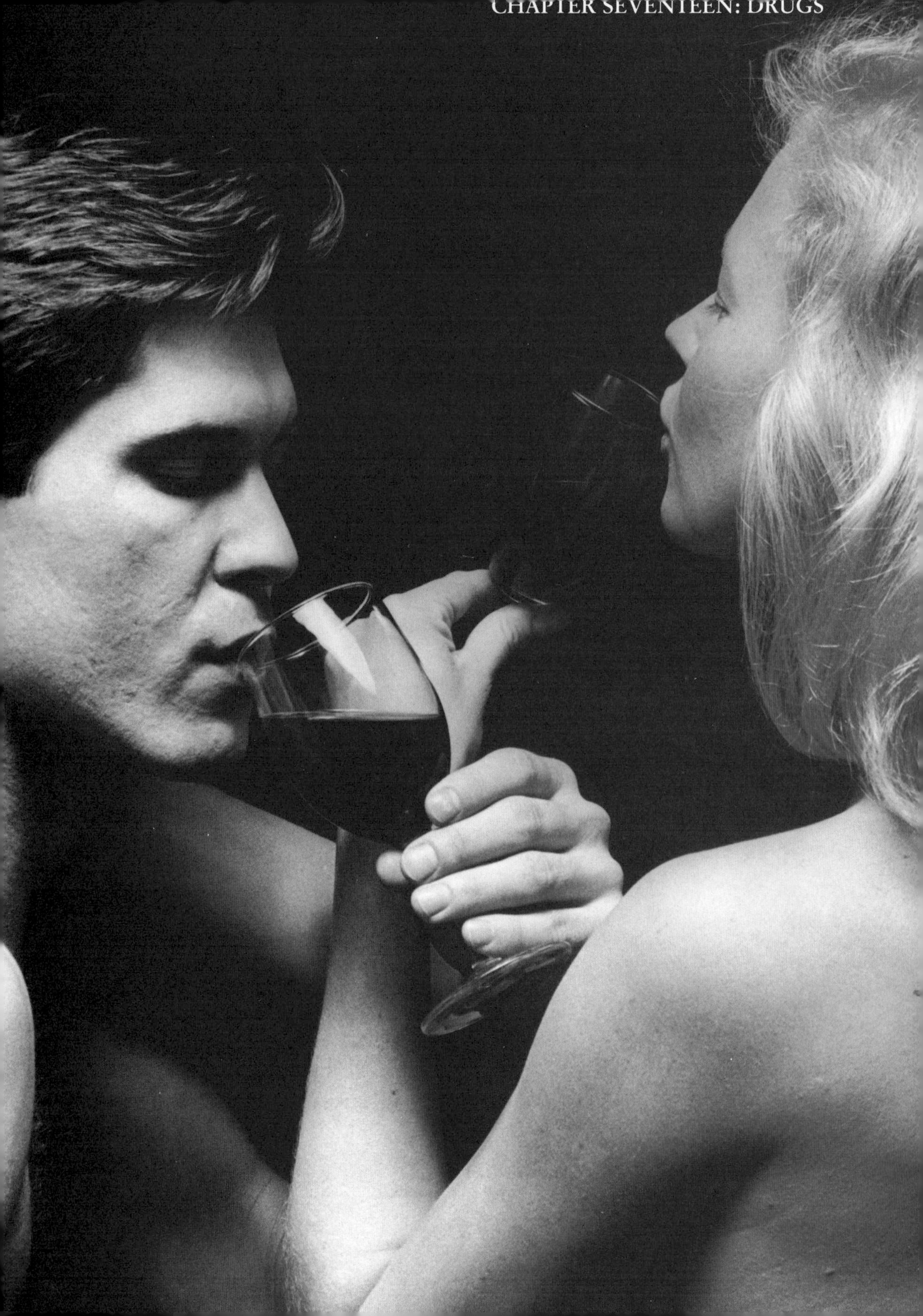

Essentially, anything which changes sensory awareness affects touch. This includes chemicals which alter one's perception, mood, or thought. Many people use drugs — legal or otherwise — as recreation aids. They're taken for a variety of reasons, including relaxing, partying, making love, and escaping.

Of course, this can be taken to extremes, and we've all heard about the horrors of drug dependence. The question of whether or not someone should use a drug is up to the individual. No one should start using some substance simply because it acts as a "touch enhancer." Moreover, many of the drugs discussed here are illegal, and some of them also have serious physical and psychological effects apart from any touch aspects. The summaries in this chapter are provided for informational purposes only and are not intended to encourage drug use.

The descriptions of each drug below are based upon both experimental studies by researchers and personal experiences of myself and others. It should be kept in mind that reactions to chemicals vary with each person. The individual personality as well as the setting intimately affect the experience. In almost all cases, it pays to be in a relaxed, hassle-free place with good company. Peaceful surroundings and trusted friends can make the difference between an exquisite experience and a real downer. Equally important, being in the wrong company under such circumstances can be dangerous and harmful.

Stimulants Stimulants include the amphetamine compounds ("speed," "ups," "pep pills," "bennies") and cocaine. (Caffeine is also a stimulant, but it's rarely used in the recreational sense we're talking about here.) These chemicals can make you feel alert, self-confident, or irritable. They increase the heart rate, raise blood pressure, and make you breathe rapidly. They kill the appetite and make it hard to relax. They can also cut down the circulation of blood to the skin, making you feel numb or cold.

None of these effects is much of a touch turn-on. Being jittery, chilly and tense is hardly a good way to enjoy your body. The drugs, of course, do stimulate your mind, which may make you more interested in bodily pleasures. However, your body simply may not perform well. This is more of a problem for men than for women. For example, because they impede circulation to the skin, stimulants make it difficult for men to sustain erections or ejaculate. And what about cocaine's legendary reputation as an aphrodisiac? There's certainly little objective evidence to back up the claims. Coke's widely perceived association with glamour, class and sinfulness provide part of the explanation. The euphoric high connected with coke might also loosen up inhibitions about the body. All in all, however, the stimulants can be considered to be basically touch turn-offs.

Depressants I include anything here which acts to depress the nervous system (except alcohol, which is discussed separately below). This includes barbiturates ("downs"), narcotics (e.g., heroin, morphine, codeine) and tranquillizers (major ones like thorazine and minor ones like librium). Generally, these drugs dull the senses, making you withdraw into your own private world. Most of them knock the sex and hunger drives for a loop, and you can expect the same for touch. In large doses, the barbiturates, narcotics and major tranquillizers turn you into a vegetable. Of course, some of them have been used to seduce women by putting them "out of it," but this can hardly be considered sense enhancement.

Small doses of minor tranquillizers like librium and valium may relax some uptight types enough to get them more attuned to their senses. This is also true of similarly calming compounds — the "hypnotics," "soporifics," and muscle relaxants. These include such drugs as dalmane, miltown and methaqualone (quaaludes). There's been a lot of press recently about quaaludes being a new "love drug," but the explanation for that is the substance's ability to break down inhibitions. All the mild tranquillizers can help a rigid, controlled person do what he really wants to do inside. Apart from that, none of the depressants do anything to enhance touch. By blunting reality and screwing up thinking, they actually make you *less* sensitive.

Alcohol Liquor first stimulates, then depresses the nervous system. In moderate amounts, it also loosens inhibitions and may break down barriers to touch. In larger amounts, it dulls reality much like the depressants. I see no harm in accompanying certain touch techniques with a glass or two of wine, unless you're doing a lot of work on the digestive track. A bottle of quality wine or champagne lends a romantic air, and the warm glow of a few drinks does seem to stimulate bodily sensitivity. Moderate amounts of beer of hard liquor produce similar effects. If you overdo it, however, you'll end up uncoordinated, awkward, and unreceptive.

Marijuana Pot heightens sensory awareness, which makes it an ideal touch-enhancer. Food seems to taste better, sounds seem clearer, and things seem nicer to touch. Pot's effect on the system is a sort of mixed bag — a bit of stimulation, depression, and even some hallucinogenic qualities. It also seems to slow down the user's sense of time, which allows him to appreciate the subleties of bodily sensations. Grass smokers commonly report that textures seem more varied, and soft objects seem more inviting to hold.

    Pot is also commonly considered to be an aphrodisiac, which can probably be explained by its tendency to intensify feelings in general. To some extent, that's all that really counts when it comes to feelings. If it *seems* as though you're feeling more, than you really *are*, because all feelings are subjective. Marijuana works well with most touch techniques, particularly such sensual ones as massage and cuddling. A joint and a glass of wine can also be enjoyable preliminaries for love-making.

Psychedelics Out of all the controversy and commentary concerning these chemicals, little mention was made of their touch potential. LSD, mescaline, psilocybin and similar compounds all have profound effects on the nervous system. Most researchers feel the drugs affect the transmission of nerve impulses, allowing many more inputs to flood the mind. This means that a lot of information which is normally suppressed freely flows into the mind. Psychedelics generally arouse the nervous system, but they also affect the emotional centers of the brain, which makes for mood-alteration.

    Physically, LSD and its cousins speed up the pulse and heart beat, and cause raised blood pressure, irregular breathing and temperature changes. Nausea and a loss of appetite can also occur. Awareness is heightened, and all the senses seem more acute which is where the touch potential comes in. The senses may seem to merge (e.g., you may "see" sound), hallucinations and perceptual distortions may occur. All sorts of buried associations pop up into consciousness. For example, looking at a woman may bring ideas of a goddess, harlot, bitch, mother and lover, one right after the other. Everything you experience seems "heavy"—profound, laden with poignant meaning.

Naturally, touch is also greatly affected. The entire skin surface seems charged with sensitivity. The air surrounding you may feel as dense as water, while immersion in water can lend the feeling of being massaged all over. Of course, this almost inflamed sensitivity markedly influences your touch perceptions with people. Simply holding a lover's hand can feel like an impending orgasm. Cuddling, massage and other intense touch experiences are deepened and enriched —*if* you get into them in the first place. The psychedelics often take you into wild, fantasy head-trips which make it almost impossible to concentrate on the body.

If, however, you're in the right mood, in the right setting, and with the right person, you can be catapulted into touch-heaven. The richness of the touch experience under LSD escapes words. Love-making takes on a primal, mythic quality, almost as though you are Adam and Eve doing it for the first time in history. Some researchers claim that psychedelic-sex can add entirely new dimensions to your sex life: "These involve awareness of more basic forms of biological processes. Subjects tend to use such extravagant-sounding phrases as 'cellular orgasm,' 'pulsating energy patterns,' 'internal fire flow,' 'melting and flowing of the entire body,' etc., in their descriptions of this experience" (T. Leary, R. Alpert and R. Metzner, quoted in the *Catalog of Sexual Consciousness*).

A psychedelic-enhanced orgasm can feel eternal and absolute, as the mind and body flow together in one vibrant pulsation. It has been described as near the top of any list of peak experiences. For example, "the orgastic ability of an individual is usually greatly increased in both male and female subjects. Sexual intercourse on the (LSD) session day can become the most powerful experience of this sort in the subject's life." (*Realms of the Human Conscious*, Stanislaus Grof, MD, quoted in the *Catalog of Sexual Consciousness*).

While LSD and its brethren definitely rate as supreme touch-enhancers, they should not be approached lightly. A trip can last eight hours or longer, and you won't be able to sleep, eat or be fully relaxed for all that time. You could get nauseous, and you could even "flip out," or lose control. These dangers are heightened because of the poor quality of much of the psychedelics sold in the streets. You have to be motivated by more than just touch-enhancement to use these drugs. Not only that, but they're scarcely available any more, legally or otherwise. The psychedelics seem to have faded along with campus protests and civil rights demonstrations. However, if you happen to be among the small group who still uses them, you may want to try touch techniques during your sessions. You'll find they mutually enhance each other. While the drugs boost your touch sensitivities, techniques like massage help relax you and take the rough edge off the drug experience.

Odds and Ends PCP ("angel dust") is a potent animal tranquillizer. It depresses the nervous system, and will depress your touch sensitivity as well.

Nitrous oxide ("laughing gas") is most commonly used by dentists to sedate patients during dental surgery. It produces a pleasant, hallucinogenic euphoria, but its anesthetic qualities make it a poor candidate for touch-enhancement.

Amyl nitrite ("poppers," "snappers") is used by medical personnel to speed up the heart beat and respiration of patients. Some people — particularly in the gay community — sniff it right before orgasm to boost the sex experience. Whether or not it enhances orgasm, the drug's intense effects on breathing and heart rate give it poor touch-potential.

Tobacco, which acts as a mild stimulant, tends to constrict blood vessels and cut off circulation. So — apart from the fact that it stales your breath, speeds up aging and can kill you through lung cancer — tobacco is also a lousy touch-stimulus.

MDA (phenylisopropylamine), a non-hallucinogenic psychedelic, had quite a reputation as an aphrodisiac in the early 70s. Studies showed that the drug enhanced emotional sensitivity, decreased anxiety, and opened communication and empathy. "MDA's effect is primarily on the emotions and body sensations, and produces a state of 'centeredness' free from the perceptual and mental distortions and hallucinations so common with LSD," reported the *Psychedelic Review* in 1969. Sounds like an ideal touch-enhancer. But that may be academic, because MDA is even scarcer now than other psychedelics.

Touch-enhancing gadgets run the gamut from the sublime to the ridiculous. They can be as cheap as a dime-store back scratcher or as dear as a professional-quality spa. Some are as simple as a carved block of wood, while others are complex systems of electronic equipment. Some are intended for solitary enjoyment, while the essense of others is socializing. A few are mainly sex enhancers, while some are basically physical therapy aids.

The point is, you're covering a wide range of territory when you talk about "touch-enhancing devices." The only thing they really have in common is that they're all non-human means to receive skin strokes. And if you think about it, you could also include being caressed by sun, surf or wind within this definition, though these can hardly be considered "devices." We'll be sticking to man-made creations in this chapter, letting nature take care of herself. It's worth mentioning, however, that a warm, sunny beach or ordinary pool offer generous sources of non-people touch stimulation. Many touch-devices rely mainly on water, and there's plenty of free water in the oceans and lakes of the world.

To some people, man's recent inventions have done nature one better. Several of the devices in this chapter provide the ultimate in body luxury. Hot tubs, spas and saunas—to mention a few—can pamper you, cleanse you to the core, relax you. Some of the sophisticated massagers may send you to touch—and even erotic—heaven. Other devices can ease sore muscles, loosen up accumulated tensions, and help you to attain internal transcendence. For the sake of simplicity, the items have been arranged alphabetically.

The Bath The old-fashioned, leisurely bath has fallen into neglect lately. The frenzied pace of modern life leads most people to the five-minute shower as the basic cleansing mode. While the shower may be quick and convenient, it doesn't offer anywhere near the benefits of full body immersion. A warm bath (temperature approximately 98 degrees) has a long tradition as a muscle relaxant and insomnia cure. "A leisurely soak in a warm tub is one of the supreme pleasures in the human experience" say Gregory and Beverly Frazier, authors of the *Bath Book*.

Everyone should enjoy being pampered with at least one unhurried bath a week. Of course, if you have the room, an attractive companion can add to the enjoyment. You can wash or massage each other's back—as well as anything else which comes to mind. But you don't need a sexy friend for hedonistic pleasures in the bath. By creating the right setting, you can turn an ordinary soak into a royal treat. Get yourself a quality soap, a bath mitt, strap or brush, and your favorite additive. Bath oils, bubble baths and milk baths are just a few ways to put magic into your bath water.

To enhance the experience even further, dim the lights, turn on some mellow music, and light a fragrant candle or incense. Downing a nice glass of wine won't hurt. Then just lie back on a bath pillow and feel those aching, tired muscles and bones being sweetly soothed by the caressing water. You'll come away clean, refreshed and relaxed.

Body Rollers There are several recently developed products which are designed to stimulate acupressure points on the body. Generally, they are carved blocks of wood which are contoured to fit certain body spots.

One of the most popular devices is the Footsie Roller, manufactured by

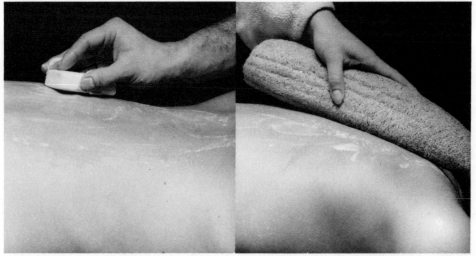

Bath pleasures, soaping up a friend   *Bath pleasures, loofah rub*

Natural Energies (P.O. Box 8010, Ann Arbor MI 48107). A nine-inch long piece of carved cherry wood, it has ridges of various widths to stimulate different parts of the sole of the foot. You sit down and roll your bare foot back and forth over the product. This is supposed to hit many of the pressure points in the foot which connect with various internal organs. I've used the Footsie Roller after running and other hard exercise, and find that it definitely soothes and invigorates tired feet. When you put solid, steady pressure on it, you can feel the foot reflexes coursing deeply throughout the body.

Another popular roller is the Ma-Roller, offered by Great Earth Healing (660 Elm St., Box Q3, Montpelier, VT 05602). A fifteen-inch piece of carved maple, it can be used anywhere on the body. However, its design works particularly well on the back. Two curved knobs in the center of the roller are separated by a two-inch recess, and the spine fits nicely into this space.

Starting from the top of the back, you move the roller inch by inch slowly down the spine, resting for a few minutes on each move. The knobs stimulate the important acupuncture meridians which lie on each side of the spine. It takes a little getting used to, but once you learn to relax and breathe deeply with the roller under your spine, the tight back muscles loosen up under the knobs' pressure. It feels much like a Shiatsu thumb-massage.

Hammacher Schlemmer Hammacher Schlemmer is not a device, but a highly regarded New York City specialty shop which stocks just about any touch-device imaginable. Their catalog contains items which are seldom seen elsewhere. Here is a partial list: thermal heat massager (you lie on it and it heats and massages your back); heating pads (all types); back-scratcher (which doubles as a shoe horn); facial bubble bath (mechanical device which uses bursting bubbles to massage face muscles); Austrian back scrubber; sack comforter (a warming, sort of sleeping bag which you can move around in); portable steam sauna bath (box-like device you get inside of); portable whirlpool bath (you attach it to the side of your bathtub); mechanical massagers (all types).

Any of these products can be ordered toll-free at (800) 228-5656; New York City residents, call (212) 937-8181. I highly recommend their catalog as a great source of gift ideas, touch-oriented or otherwise.

Health Clubs and Resorts Once again, these aren't devices, but they are often the only affordable source of many expensive touch-items. They are often referred to as spas, a term derived from the mineral springs in Spa, Belgium. At one time, only resorts built around mineral springs were considered spas, but now the word is used for any luxury resort or hotel. To further confuse things, a specific type of whirlpool equipment (discussed below) is also called a spa.

Such health clubs and resorts often contain steam baths, saunas, and whirlpool devices. You may also be able to take advantage of exercise equipment, masseurs, and regular swimming pools. The resorts usually specialize in daily regimens for particular problems — i.e., dieting. Each resort or club varies, so you should check the facilities before signing up.

Hot Tubs The hot tub is making a big splash all across America. After taking off in California, it's now creating a lot of waves in the Northeast. Basically, the hot tub is a large, round, wooden tub filled with hot water (usually between 102 and 105 degrees). The most common material used is redwood. Tubs come in various widths and depths, depending upon how many people you want to entertain.

The tubs can be set up freestanding, or recessed into a wooden deck, patio, or even into the ground. A heater and filtration system keeps the water warm and swirling. Since the body retains the temperature for about twenty minutes after getting out, tubs are often put up outside, even in wintry areas. There's something especially delightful about being able to thumb your nose at frigid, snowy weather while you bask in the warm glow of the heated tub. Many hot tub aficionados add refinements, like whirlpool jets which provide powerful streams of water for therapeutic hydro-massage.

The big payoff from a hot tub is relaxation, but it's also becoming a chic new form of socializing. People are increasingly using the tubs to party, get high, or even close business deals. They can also serve as great seduction settings, especially if you don't let the water temperature get too high (which seems to boil away the erotic urges). There are usually benches built into the tub to seat guests, and you can add a bevy of accessories: extra seats, towel racks, beverage shelves, covers and steps. Hot tub folklore has it that chilled white wine and pot go particularly well with the experience. Mellow music and dim, candle lighting also add to the mood. All in all, any self-respecting hedonist would have to rate the hot tub high!

Jacuzzi Whirlpool Spas Jacuzzi is a brand name for a whirlpool-jet system, but most people use the word to refer to *any* such system. Unlike hot tubs, spas are usually constructed of plastic (fiber glass or acrylic). Most spas are rectangular or square-shaped, and late models are furnished with sculptured seats and lounges. They come in all sizes, and basically produce the same hot water action as the whirlpool-equipped hot tub. The main difference is in materials used: the hot tub has a rustic, natural feel, while the spa has a clean, modern look. Hot tubs, however, hold up better over time.

"Hydro-massage" is the key benefit of whirlpool treatments. The hot, moving water created by the Jacuzzi jets sends streams of soothing bubbles against the body. Hydro-massage is effective for many ailments, including sprains, strains, bone injuries, arthritis, spasms, bruises, and circulation problems. Even perfectly healthy people will love the calming, rejuvenating effects.

Whirlpool spas are generally constructed to hold a half-dozen or more people, and are often built into sun decks, patios, or lounges. Smaller spas can be found in oversized, luxury bathrooms. Some companies manufacture whirlpool bathtubs and shower installations. You can even buy a portable jacuzzi which attaches to the side of your regular tub. No matter in what form you experience it, the jacuzzi is a treat for tired muscles!

Mechanical Massagers This is an area surrounded by controversy, because many of these products have obvious sex angles. The explicitly sex-oriented devices include battery-powered plastic vibrators, which are generally phallus-shaped. Despite the suggestive contours, these noisy little items are not usually used for penetration, but for clitoral stimulation. Nonetheless, they *are* capable of more than that. The vibrating action can be used for localized muscle stimulation anywhere, though the batteries would run pretty low if you tried anything like a full-body massage.

A big step up from the portable vibrators is the "personal massager." The typical unit is shaped like a large phone handle with only one head. The massager's head usually can be fitted with a number of attachments designed to do different things. Some of the more common ones include face, scalp and general body stimulators. The personal massager business is in a real boom, with new manufacturers entering the field continuously. At this point, it's possible to purchase these products in such respectable department stores as Macy's and Gimbel's in New York. That wasn't always the case.

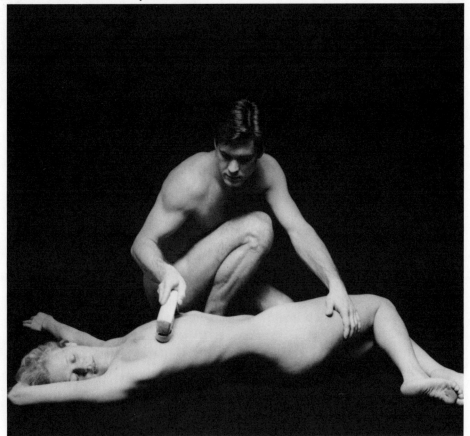

Mechanical massage

The pioneer in the personal massager field is Richard J. "Tex" Williams, founder of the Sensory Research Corp., which makes the Prelude 3 unit. For years the company's sales were limited to mail orders, and it wasn't until 1978 that Preludes began selling at retail. Part of the problem was that Preludes used to be sold with a sexual slant, the promo material stressing the vibrator's effectiveness for female orgasms. And Williams was quite right on that score: the vibrating action on the clitoris often works where nothing else will.

Eventually, Prelude's slant shifted to that of a general body massager. In July, 1978, it gained legitimacy by being introduced at the National Housewares Exposition. Coincidentally, the Prelude 3 was selected as the "critics' choice" in the March, 1978 *Playboy* "Sex Aids Road Test." The reason for that is simple: the unit comes with an additional attachment, a knob-shaped spot stimulator specifically designed for the clitoris.

The general body attachment works pretty well on the penis. If that's not enough, however, a special phallus-stimulator attachment, the "Pleasure Dome," can be purchased separately.

The sales success Prelude 3 has achieved since going retail has encouraged many big-name outfits to enter the field. Clairol, Conair, Norelco and Water Pik have all recently joined the fray. Hitachi, Pollenex and Oster are also major manufacturers. Most of the companies ignore the sex aspect in favor of a health and beauty angle. However, many of the new products have a spot stimulator which looks suspiciously like the Prelude 3 clitoral attachment.

Clairol's twin-speed, bright green Body Language unit has four attachments: facial, scalp, hip and thigh, and general body. The hip and thigh, and the facial attachment, can be used to apply creams and lotions over those areas. The device has a strong, though quiet, motor and is definitely capable of giving a deep, soothing full-body massage. If you're creative enough, you shouldn't have too much trouble figuring out how to stimulate yourself any way you want. Oster's Vibra-Massage, Norelco's Feelin' Good, and Conair's Massage Works II all seem essentially similar to the Clairol product. However, they all contain that additional knoblike "spot" attachment for "localized stimulation."

Oster and Pollenex are the largest massage manufacturers, and they have a number of different product lines. Both of them produce a "Swedish-style" massager which straps to the back of your hand. This type allows your own skin to administer the vibrating action, and is often preferred by professional masseurs. Oster and Pollenex also make "cushion massagers" for the back. Water Pik's new LeBody massager and several Oster models combine vibrating attachments with heat.

Another popular massage style is the "wand" type with an extra long handle for reaching hard-to-get spots. Hitachi, Conair and Prelude all have versions of this. Hitachi also puts out a unique "twin-head" massager which lets you reach *both* sides of the neck, shoulders, arms, etc. Eve's Garden, an innovative sex-products boutique in New York, claims that the Hitachi twin-head unit is also perfectly suited for penile stimulation. "Recommended by a male sexologist to be ideally suited for ultimate satisfaction for men," says the Eve's Garden catalog.

If you're interested in top-of-the-line body massagers, you won't want to overlook the Massage Vibrator made by the Battle Creek Equipment Co. of

Michigan. The hand-held unit plugs into an external rheostat for precision speed control, and has a variety of cup, ball and sponge applicators for all types of pressure. A unique U-shaped applicator is designed for the spinal area. The full package costs almost $600.

Another recent development in mechanical massagers is the foot-massager. Rest your feet inside Clairol's Foot Fixer on twin arch rests which contain dozens of tiny rubber protrusions called "Vibra-fingers." When you turn the massage switch on, these things vibrate, causing a gentle, soothing effect. You can use the device dry, or fill it with hot water with your favorite skin conditioner, oils, or minerals. Another switch will maintain the water temperature. All in all, a pretty handy way to soothe sore feet. Other manufacturers make different types of foot massagers, including Oster's Hydro-Lax and Whirlpool Massager.

Saunas and Steam Baths The sauna experience is catching on in the U.S. It is a major social institution in Finland, which — with almost a million sauna units — registers one for every five Finns. There are two basic phases to the sauna experience: relaxing in a superheated wooden room (180 to 200 degrees), followed by a plunge into cold water or snow. The wooden building (often redwood) can be closet-sized or a lot larger. It is heated by a gas, wood, or electric stove. Hot stones set on top of the stove are periodically splashed with water to produce steam. The sauna room generally contains two tiers of benches for sitting or reclining.

Sauna, like all steam baths, causes profuse sweating, which thoroughly opens and cleanses the pores. Besides deep cleaning, sauna benefits include improved circulation, profound relaxation, and a "heightened tactile sense which linger(s) for several hours afterward," according to Johnson and Miller, authors of the *Sauna Book*. True sauna aficionadoes follow this cycle: First, they relax and get heated up in the sauna room. Then they cool off rapidly, either by showering, having a bucket of cold water poured over them, diving into a cold tank or lake, or rolling in a snowbank. The snowdive may seem a bit drastic to some, but Johnson and Miller describe it as "one of sauna's greatest pleasures, no matter how terrifying it sounds." The last part of the cycle is rest.

This routine is repeated as many times as feels good, with three being about right for the average individual. In the last round, many sauna devotees whisk themselves with birch or oak branches. This is followed by a soap and water wash in the hotroom or outside, then one last cold plunge and a gradual cooling off. The sauna experience can be enhanced by putting a few drops of scented oil in a water-filled can on the stove. This will fill the room with an air of wintergreen, peppermint, or whatever scent you choose. After finishing the sauna, the best thing to do is just lie back and enjoy the deep relaxation. If you can coax someone to massage you, all the better.

Sauna units can be constructed indoors or outdoors. Some people erect them in basements. Many manufacturers sell prefabricated structures, or you could build your own. The *Sauna Book* — the last word on the subject — contains a lengthy list of manufacturers throughout the U.S., as well as complete building plans for the do-it-yourselfer.

Unlike saunas, which use dry heat, the Turkish-style steam bath relies on high humidity. However, its relaxing and cleansing effects are similar. Modern steam baths are usually large, public rooms in health clubs or resorts. However, some manufacturers are producing portable steam generators which turn your stall shower into a private steam bath. These generators are mini-versions of the ones used in commercial outlets. They're small enough to install under the bathroom sink, and contain electrical timers, so you can set the duration of your steam exposure in advance. After the steam bath, you simply turn on the shower to rinse yourself off.

Portable steam generators include the "Steamist" and Amerec Corporation's "Steamer." Water Pik also makes a shower-attachment steamer, which is described below.

Shower Attachments A popular new addition to many households is the pulsating shower spray. The top seller in the line is the Shower Massage by Teledyne Water Pik. It comes in two models, a hand-held movable unit and a wall mount. You simply replace your old shower fixture with the Shower Massage.

The dials on the unit allow you to choose different combinations of ordinary spray and massage pulsations. You can set it for pure spray, massage only, or spray/massage (which is particularly good for facial massage). You can also control the speed of the massage action, from slow-mo to as many as 9000 pulsations per minute. Besides just feeling food, the Shower Massage can help relieve such ailments as backaches, muscle spasms, neck pains and minor arthritic pains. It's particularly useful after heavy exercise, when you get the benefits of soothing massage and a cleansing shower. Conair's Waterfingers and Pollenex's Dial Massage are similar water pulsators.

A recent innovation in this area is the Steamer shower massage by Water Pik. This combines the benefits of the shower massage with a new feature: steam generation. The unit has a special "steam position" which shuts off the shower (so you don't get scalded by dripping hot water). You turn the water temperature up as high as it will go, and a steam mist is emitted from the underside of the device. In five to ten minutes, you should get the full effects of the steam atmosphere. It won't be as intense as a commercial steam bath, but you can also stay in it longer.

The Steamer unit comes with a shower curtain extension to cut down on drafts. The company suggests using a small stool in the shower so you can sit and relax. And as I mentioned above, deep relaxation is the key benefit to be gained from saunas and steam baths. The Steamer shower massager certainly sounds like an ideal sensory toy: you can steam, massage and shower yourself all in one sitting, without even needing another person to give you a hand. It could turn recluses into hedonists!

Skin Care Products Many beauty treatments offer interesting sensory experiences. This is particularly true for men, who rarely try them. For example, the facials which women commonly receive in beauty salons can make you feel quite pampered. In the *Total Beauty Catalog*, K. T. Maclay describes a luxurious salon facial: "the assistant steamed my skin with a fragrant herbal tea, cleaned my pores (by hand), applied three different types of masques (which smelled strongly of camphor and other good stuff) and sent me out into the world feeling (and looking) 200 percent better."

Of course, you can always apply your own facial masques. There are three basic types: creams (which moisturize); pastes (which harden on the skin, clean-

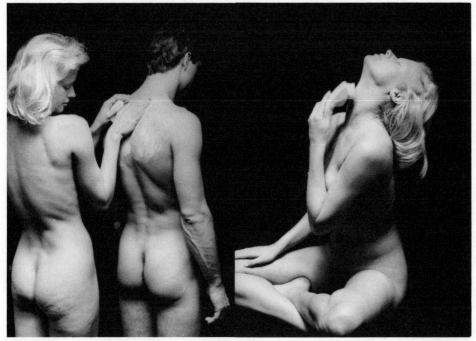

Shower play   *Skin care delight*

ing pores and correcting oil imbalances); peel-offs (which tighten the skin and pull off dirt, blemishes and other impurities). The nice thing about these products is that they combine interesting sensory stimuli with healthy skin care. There are many commercial masque preparations available at any cosmetics counter. The *Bath Book* includes some intriguing do-it-yourself masque recipes, like "avocado and yogurt" and "orange and honey."

For those preferring the ease of electronics, there are skin care gadgets on the market. For example, Conair's Shape-Ups device does both facial and nail work. Several attachments are for manicures and pedicures, and the soft nylon brush and moisturizer are for facials. But the latest word in beauty technology appears to be the Skinvention by Clairol. It has four attachments: a pumice stone for smoothing callouses, skin buffer for removing dead, flaky skin, facial brush and body brush for cleaning and moisturizing those areas. The battery-operated unit is fully immersible, so you can use it in the bath or shower. While skin cleansing and conditioning are the key benefits of the Skinvention, Clairol PR person Nancy Coleman points out that some people use it simply because it makes them feel good.

Tranquility Tanks The tranquility tank is the brainchild of Dr. John Lilly, a neurophysiologist who has done notable research in many areas, including human consciousness and dolphin communication. The tank is an eight-by-four-by-four foot chamber which contains a shallow 20 percent epsom saltwater solution. Your body will float like a cork in this. The water is heated to skin temperature, and the tank is kept as sound and vibration-proof as possible. When you close the lid, you'll be in the blackest black you'll ever see — or *not* see.

The basic idea behind the tank is to limit your sensory stimuli to the lowest conceivable level. About the only thing that's really left is the slight tactile sensation of the water against your skin. In the quarter-decade or so since Lilly has been experimenting with the concept, he's found that these conditions produce the most profound relaxation possible. The weightless effect eliminates most of the muscular tension expended dealing with the usual gravity field. For perhaps the first time, the mind is free to think without physical fatigue affecting it. "In summary, then, the tank experience is a very refreshing one, a resting one," Lilly says in his book, *Deep Self*.

But there's more to it than just rest. In total sensory isolation, you become deeply absorbed in internal processes. You become intimately aware of internal sounds like heart beat, breathing and gut gurgles. Your eyes may manufacture visual patterns out of the darkness. You may also journey far into the inner recesses of your own psyche. *Deep Self* records the personal tank experiences of dozens of people, including many celebrities. The anecdotes include out-of-the-body experiences, mystical states of mind, and profound psychological insights. The more tank sessions you have, the further you travel into these inner realms.

Some observers feel that the tranquility tank will one day become a hot trend. There are presently only two manufacturers, the Samadhi Tank Co. of L.A., and the Denver-based Float to Relax Inc. The tank set-up, complete with filtration system, sells for about $2000. *Deep Self* includes building plans for do-it-yourself-ers. If neither buying nor building appeals to you, there's one other alternative. Both companies have been franchising the tanks to different centers throughout the country. For fifteen to twenty-five bucks, you get an hour's worth of tank time.

Samadhi tanks can be found at the Samadhi Spa in Beverly Hills and at the East-West Center for Holistic Health in New York. Float to Relax tanks are located at Nautilus-plus, a Southern California health club chain, and at the company's newly-opened center in Philadelphia. Both manufacturers have ambitious plans to expand the number of tank outlets, so it may not be too long before they reach your town.

An hour in the tank at the East-West Center for Holistic Health in New York easily lived up to its billing. It was ecstatic to comfortably float on the saltwater surface. At first, physical impressions predominate, as you become intensely attuned to your breathing, heart beat and stomach growls. However, the psychological effects deepen with each passing moment, and you eventually feel like a disembodied mind spinning freely in outer space. The anti-gravity sensation sends you into such a relaxed state that it is difficult getting up afterward. Your consciousness seems to float in the twilight zone between wakefulness and sleep. A very spacy experience, and highly recommended.

Waterbeds The waterbed is a marvelous sensory enhancer, as anyone who's ever tried one can attest to. Unlike ordinary mattresses, the waterbed provides uniform support for the entire body. It conforms to your exact shape, fitting the various contours of your frame. A regular flat bed can cause pressure on your body's bulges, but the waterbed cradles you completely. This "flotation support" permits a soothing, deep relaxation. The undulating effect of the bed also enhances sex-play. Your bodies seem to merge and flow with the water's rhythm. Not everyone prefers to sleep on waterbeds, but they're certainly nice to have around for some sensual fun.

A good waterbed set comes equipped with a mattress (with solid seam construction to prevent leaks), safety liner, frame, platform, pedestal and heater.

Some new types of construction virtually eliminate wave ripples. Your local waterbed dealer can show you several convenient sizes, and many attractive frame designs. A fully assembled king size waterbed weighs about 33 lbs. per square foot when filled. This is less than your refrigerator, and should be no problem for any structure which complies with the usual building codes.

Womb-Box The "womb-box" (or "cuddle —" or "massage-box") is not a product on the market, but the product of this author's mind. It would be therapeutic to experience a nurturing, full-body enclosure. Cuddling with other people just doesn't capture the total envelopment of the womb experience. If we all had womb-boxes to which we could occasionally retreat, we might go back out into the world with renewed vigor and adventuresomeness.

Envision a large, velvety-lined box with pillowy soft insides. A person could snuggle inside, close the door, and be free from all worldly cares. With today's electronic wizardry, it shouldn't be that difficult to make the pillowy insides move for a massage-like effect. A similar technology may already be employed in the "cushion massagers" now on the market. Modern technology should also be able to overcome whatever problems breathing in such a device would present. Are any manufacturers listening?

CHAPTER NINETEEN: THE SENSUOUS ART OF CUDDLING

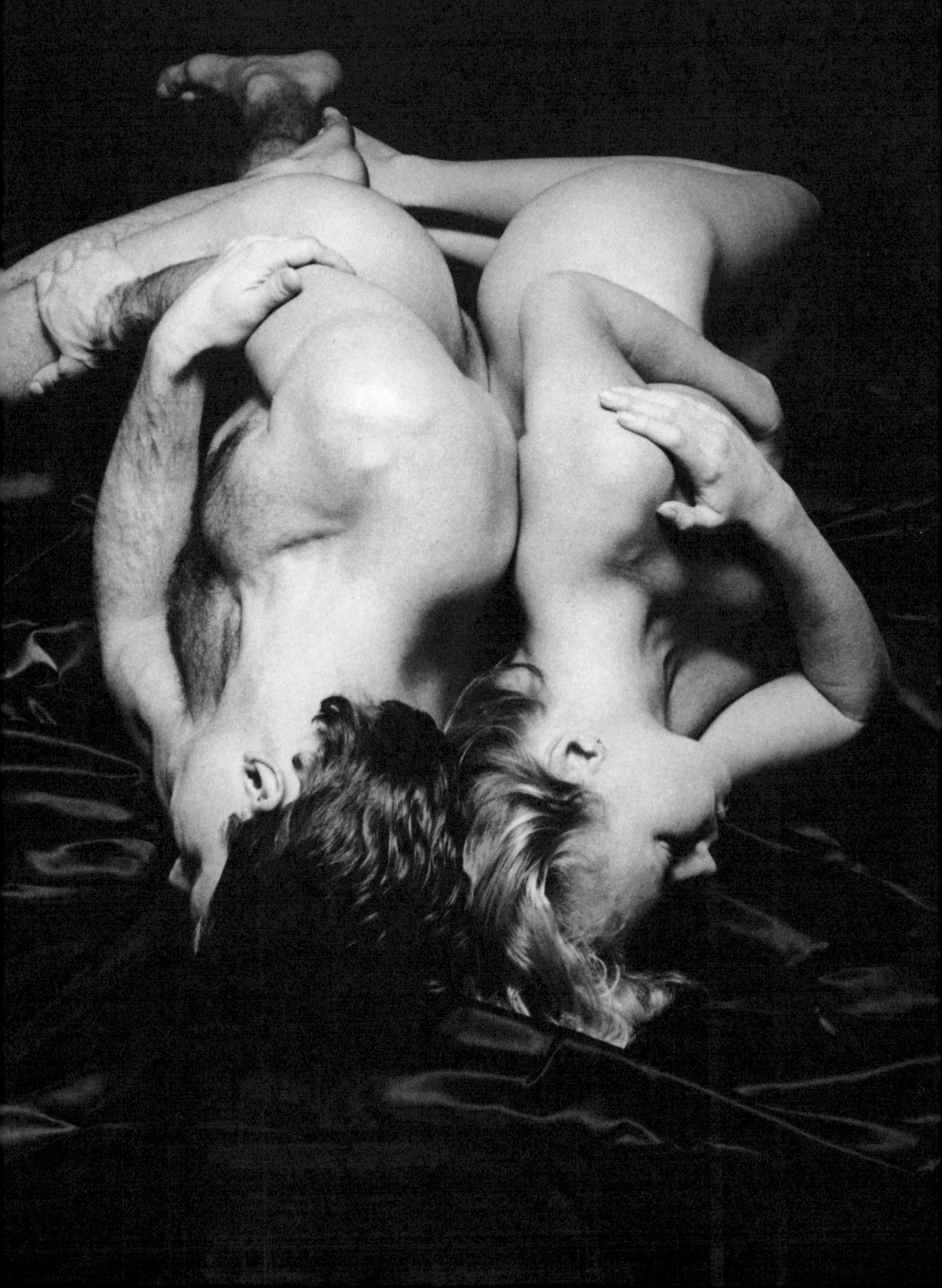

In Chapter Four, we looked at the "cuddle-crave," the deep-seated drive we all have to hold and be held. Consider now a technique which directly caters to the cuddle crave. The "sensuous art of cuddling" consists of a series of comfortable, relaxed positions which maximize body contact with your partner. It's about as intimate as any act short of sex, so you'll likely limit it to particularly close touch-partners. Intense cuddling really works well with someone you're chemically attuned to. In sex, it's possible to get wrapped up in a private, genital focus. But in cuddling, you really have to be into your partner's *entire body*, head to toe.

Cuddling can work better than sex to test out how attracted you are to someone. Many different kinds of people can fire up fleeting sex feelings, but cuddling requires more. You must be motivated to *really* explore your partner in the fullest sensual way: to press him (or her) tightly all over, to rub his body all over your own, to love every nook and cranny of his physical self. But if you do find a compatible cuddle-partner, you'll truly experience sensual bliss. Cuddling is one of the most intense touch experiences you can have with another person.

It also has numerous other benefits. It's quite effective in easing the stresses of modern, hectic life. When you press every inch of your body snuggly against the soothing flesh of your partner, tension just melts away. It's almost as if nervous energy becomes absorbed through his skin. A definite energy exchange occurs in cuddling, with the rough edges in each partner's energy level smoothed out. A tense person will be mellowed, while a tired person will feel refreshed. Cuddling is particularly effective as a relaxation aid. Before bedtime, it works wonders with insomnia. A round of cuddling can have you drifting into a blissed-out peacefulness, with cares just flowing off into your partner's flesh.

Cuddling is also nice immediately after waking up. It helps you start the day more refreshed and relaxed. Other good times are after work, exercise, or any activity which causes tension or strain. Cuddling is also a perfect complement to love-making. Before sex, it slowly stimulates you to full intensity. After sex, it provides an exquisite relaxation which easily leads to peaceful sleep. But unlike sex, which is governed by natural rhythms, cuddling has no beginning or end. You engage in it until your body is loose, limp, and rested.

When you first try cuddling, you may be a bit unclear about what you're supposed to get out of it. But if you practice it regularly, you'll develop a positive craving for it. A "cuddle drive" will take its place among the rest of your human needs. It was actually there all along, but you may not have recognized it. An overwhelming number of women questioned for the *Hite Report* on female sexuality expressed the desire for more intimate, non-sexual body contact, and cuddling is an ideal way to attain it. If you cuddle with a partner regularly, you'll also find a special intimacy developing between you. It's the "cuddle bond," and it involves unique ties of trust and acceptance. It can feel as though your bodies are never totally separate, even when you're far apart. As mentioned in Chapter Four, sex therapists often use cuddling to build confidence and rapport in couples who are having sex problems. It helps restore trust and release inhibitions.

To be fully effective as a trust-builder, cuddling requires a certain attitude of mind. Unlike sex, sports, work, or almost all of our activities, there's nothing to be "accomplished" or "done" when cuddling. We are a society of doers, and we always seem to be striving for some end. Unfortunately, sex technique books often play right into this, promising us more things we can "achieve" in bed. But there is no "cuddle-orgasm." If cuddling has any goal at all, it is just to *be with your partner's body:* to leisurely bathe in the luxury of your senses with no other purpose in mind. If you cuddle with this attitude, it'll become a peaceful, nonthreatening oasis in the middle of the maddening hustle-bustle of daily life.

The Cuddle Positions All the positions should be done nude in bed. While a hearty hug can be pleasant in any state of dress, clothing would be an unthinkable hindrance to intimate cuddling. Also, it's preferable for the woman to lie to the man's left. In yoga, the yin (female) side is on the left, while the yang (male) side is on the right. The two complement each other, the yin side being dark and absorbing, the yang side light and radiant. This yin/yang energy balance is best maintained with the female on the left side of the male in all the postures.

Position One This is the most sexually stimulating position, and can be used as a form of sexual foreplay. Both partners lie side by side facing each other, female to the left, male to the right. The man puts his left arm underneath his lover's neck (rest the arm at the gap between the end of a pillow and her shoulder to avoid cramping). This position permits the entire front halves of your bodies to touch — your face, chest, thighs, genitals and feet will all be in direct contact with your lover's, depending upon the difference in your body height.

It will be most comfortable for the woman to keep her right arm at her side, but she can still hug her partner with her left arm. The male can use his right arm to hug her. You'll undoubtedly feel a mild electric-like current pass between your bodies, especially where your genitals touch. Because of this, Position One cannot be recommended for relaxation.

As you hold your lover close, you'll feel an urge to rub against each other and press different body parts together. The male will especially enjoy rubbing his nose over the creamy smooth flesh of the female's face, neck and shoulders. He can hold her even more intimately by curling his right leg over her hips and left leg. She'll find it soothing to snuggle against his chest, particularly if he's big-chested. This front-to-front posture is also quite conducive for kissing and light nibbling.

Position Two The male lies on his back, while the female rests facing him on her right side. His left arm remains under her neck like Position One, but it's now much easier for her to rest her head on his chest. Position Two allows the female to be much more active than the male. While he relaxes, she can wrap her left thigh over his body and hug him close with her left arm. The full front of his body will be easily available for her caresses.

Position Three This is the reverse of Position Two, with the woman on her back, while the man lies facing her on his left side. He can put his left arm under her neck, or snuggle his head on her right breast. She can keep her right arm by her side, or put it around his neck. This position allows the male to explore his partner's full front with his right hand. He can also extend his right arm so it fits snuggly under her breasts. He'll really enjoy resting his right leg on the flesh of her tummy. This must be done gently and with light pressure, because the female abdomen can be quite delicate. The velvety surface of her belly will feel quite nice against the sensitive skin of his thigh. He can also nudge his right thigh against her hips and thighs. For a highly erotic sensation, it can be nestled between her legs, where it'll be engulfed in the softness of the insides of her thighs. Position Three also allows the male to kiss, nibble or rub his nose on his lover's hair, face, neck and shoulder.

Position Four Here, the woman lies on her left side with her back towards her lover. He lies parallel behind her on his left side, with his left arm again supporting her neck. This is the "spoon" position many couples prefer to sleep in. It's one of the most relaxing of all the cuddle poses. Both partners experience different, but equally delicious sensations. She'll delight in the serene sense of security caused by the warm feeling of the man cuddled around her back. He'll find his excess energy and tension absorbed by the exquisite softness of her hair, buttocks and thighs.

Position Four is my personal favorite. The natural contours of the female frame provide soothing resting places to nudge against. Long, silky hair on the female really enhances the effect. This sweet feminine fur feels enchanting on the face, chest and shoulders. Enjoy burying your nose in your partner's hair. Lift the hair to expose the nape of the neck, shoulders and upper back. These delicious spots openly invite sniffing, kissing and playful nibbles. You can also reach sensitive earlobes here.

A convenient way to maintain a secure hold on her body is to drape the right arm over her front and nestle it between her breasts. The experience of your arms engulfed in her fleshy globes can be quite pleasurable for both of you, but it also makes snuggling very simple. You can hug her tightly with both arms, left underneath and right around the front. In Position Four, the complementary curves of the male front and the female back permit the maximum possible intimate contact. The male midsection flares out right where the small of her back slopes in. Her curving buttocks provides a snug fit for his flat, straight hips. This contact is a most enchanting experience, your hips and genital area nestled warmly against her soft, fleshy buttocks.

The man will enjoy the sensation of the smooth backs of her thighs against the fronts of his legs. You can also cuddle your feet together: The male extends his until they reach the silky undersides of hers. He can also nudge his toes underneath hers. A faint energy exchange will flow between your feet, which enhances relaxation.

After experiencing maximum contact in this position, stop moving and just rest. Your bodies will grow increasingly limp (assuming you're already sexually sated) as you securely heap against each other. The posture can be quite conducive for drifting off into deep sleep. Before becoming too drowsy, however, the man should remove his left arm from under her neck to avoid cramping while he dozes.

Position Five This is the reverse of Position Four. Both partners lie on their right sides, the male's back towards the female's front. Some women will find it comfortable to slip the right arm under the man's neck. However, most females will not take this pressure well, so she may want to keep her right arm at her side.

Like Position Four, the female can drape her left arm over the front of her lover to securely hold him. But in other respects, you can't directly follow Position Four's pattern. Since most women are smaller than their men, they won't be able to wrap around the male body from head to toe. Instead, the female will switch her attentions between the top and bottom portions of his body. At first, she should snuggle up against the top of his back. Then she can rub her nose around the nape of his neck, lightly sniffing all over. You'll both find this exciting—the sensitive tactile and olfactory nerves in her nose will be stimulated, while he'll enjoy a peaceful serenity from her rhythmic stroking.

The feeling of female breasts snuggled tightly against the male back is also gratifying. Equally pleasant is the sensation of her supple tummy against the small of the back. The skin of his buttocks will tingle to the lusty touch of her pubic area. If she moves down a bit, he'll feel her silky thighs against the back of his legs. She can also nestle the tops of her feet against the undersides of his for a pleasurable bout of foot-cuddling.

All in all, this position is one of the most restful for the man. Having a soft, nurturing woman cuddled closely behind you makes you serene and secure, a wonderful way to drop off to sleep.

Position Six Both partners lie back-to-back as snuggly as possible. Since your hands will face away from your partner's body, not too much stroking will occur. But that doesn't matter, because the back is a marvelously sensitive, but woefully neglected, touch-spot. This back-to-back posture permits the purest relaxation. You'll both feel secure and serene, and may find this a nice way to share your sleep-time.

Performed in order, Positions One to Six form a full cuddle cycle which moves from maximum stimulation to deepest relaxation. The postures become less erotic and more restful as you move along, until you reach the pure peace of Position Six. By following the cycle, you'll totally sate the cuddle craves throughout your body, and end up in a perfect state for deep sleep. The One to Six cycle, therefore, works particularly well as your final bedtime activity of the night. Naturally, if you're not into napping you can mix and match the positions any way that fits your mood. Following are several more positions to suit your fancy.

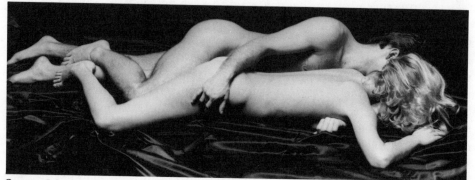

Position Seven Many women like this variation. The male lies on his left side, while the woman is prone on her tummy. He supports her neck by resting it on his left arm. His right hand will be free to pet, stroke and massage her entire back. If he wants to relax rather than stimulate her, he should lean over against her body, so his right side rests lightly on top of her back (remember, easy on the pressure). Then he can loop his right arm securely around her back. He can also rest his right thigh on her buttocks. The lower part of his leg will wind up between her legs, tightly trapped inside the creamy warmth of her thighs. This simultaneous contact with her fleshy bottom and tender thighs can be a real touch-high.

Positions Eight Through Eleven These all involve one person lying on top of the other. The key benefit here is that you add body weight to the cuddle sensation. This can be deeply nurturing for both of you. Of course, you don't want to go overboard and crush your lover. If you're the heavier one, distribute your weight evenly over your elbows, knees and trunk. Give your partner as much pressure as desired, and shift any excess weight to your elbows and knees.

Position Eight has the man lying directly on top of the woman, pretty much like the "missionary position" in sex. Since you'll be making intense front-to-front contact, this can be highly stimulating sexually. The male should hold the pose as long as his lover wants him to. Position Nine is the exact reverse, with the man underneath this time.

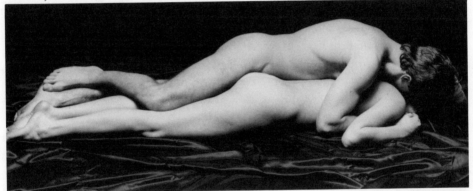

Position Ten has the woman lying on her tummy while the man rests on her back. Men will enjoy the same feeling of the female back as they did in Position Four. Long hair on the woman is an added attraction here. The female will experience deep, soothing pressure. Position Eleven reverses this, with the man on his stomach and the woman on top. He'll delight in the feminine sweetness pressing all over his back. A few women, by the way, find lying on the male buttocks sexually stimulating. Some can even come by vigorously rubbing their genitals over the area.

Positions Eight to Eleven must be attempted with common sense. You must be cautious about using excess weight. NEVER climb on a partner who has serious bone problems. On the other hand, don't be overly ginger. A varied sexual repertoire will employ similar postures, and few lovers get crushed. Just be sensitive to the weight your partner desires. These deep-pressure cuddle positions really touch you to the core. They require little energy and effort, but they fully satisfy the most intimate levels of your cuddle craves.

Sleep-Cuddling Touching someone while you sleep can be quite reassuring for both of you. We have already mentioned many cuddle postures which are quite nice to fall asleep in. However, it doesn't really matter whether you're wrapped all around each other or simply touching in one spot. The main thing is to make continued contact somewhere.

We enter a totally separate reality in slumber, a reality often accompanied by bizarre, scary dreams. We're quite vulnerable, so it makes sense to bed down with someone who makes you feel secure. Making contact with a trusted friend while sleeping seems to "ground" you. The other person is a warm, reassuring presence. There may be some sort of psychic energy exchange during sleep-cuddling. Occult teachings maintain that the "astral body" leaves the physical self during sleep, which is supposed to explain the common dream of flying. Perhaps sleep-cuddling links each partner's astral body, making them less likely to go astray.

Whatever the case, you'll sometimes find that regular sleep-cuddling with a partner will produce parallel dreams. They can even continue when you're sleeping apart. But parallel dreams aside, sustained sleep-touching with someone causes a special closeness between you. A subconscious connection, the "sleep bond," forms and some part of your subconscious remembers that warm nighttime linking of your bodies. Once you find someone with whom you're comfortable enough to sleep-cuddle, you'll never feel quite the same about solitary slumber. If you're a loner-type, however, there is an alternative. You can form a sleep bond with a pet. Person or pet, the main ingredient is total trust. Since the process is essentially subconscious, it makes little difference whether your sleep-cuddler is man or beast. As long as you feel completely secure with the contact, you'll be able to form a comfortable sleep bond.

Pet Cuddling Cuddling works well with pets, particularly if you teach them from the baby stage. Cats rate highly as sensual cuddlers; they're soft to snuggle against. Cuddle-raised cats are very down to earth, not aloof at all. Sensual, beautiful creatures, they make wonderful miniature touch companions. Rubbing your nose on their soft, sweetly-scented coats feels much like the silkly smoothness of a woman's long hair.

CHAPTER TWENTY: THE ART OF EROTIC TICKLING

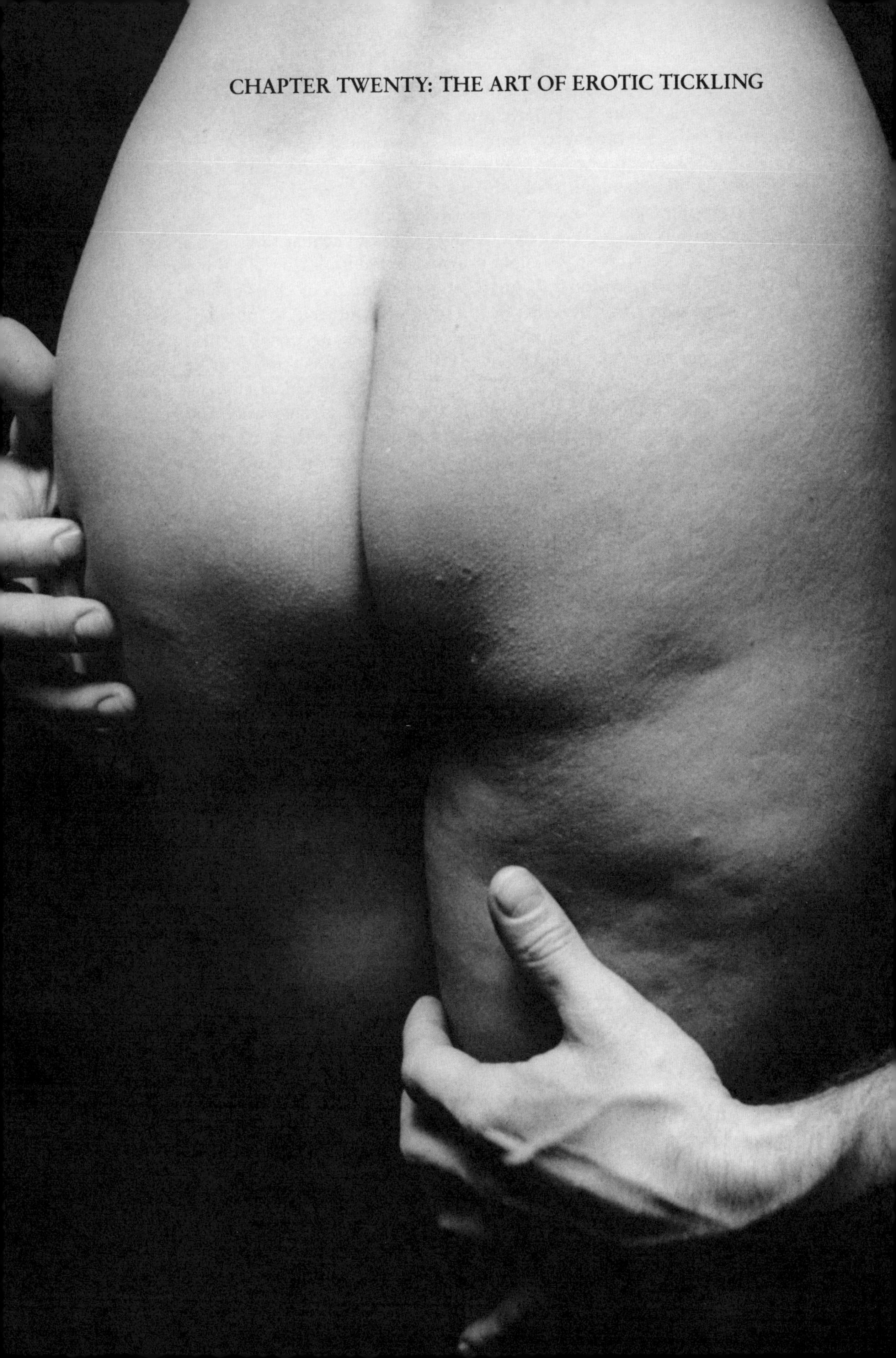

Ticklishness is one aspect of the touch experience which is almost totally overlooked. Entire books have been written about sex, sensuality and human contact without the subject being mentioned once. Most people consider it to be child's play and no more. But creative touch-partners can find a world of stimulation in tickling. It may be the sole touch sensation which mixes light-hearted fun, aggression, and erotic titillation.

And just what is the tickle-response? It's a peculiar tingling reaction our skin has to certain types of light touch. We find it pleasurable, but our bodies also shudder to avoid it. We may get goose-bumps or chills down the spine. There seems to be little scientific interest in the subject. As mentioned early in this book, some psychologists consider it a warning signal to alert us when vulnerable spots are being invaded. There does seem to be an element of aggression in the experience. When someone we don't like tickles us, we react with hostile indignation. But when a lover does it, aggression merges with affection, which gives tickling its unique creative power: a harmless safety-valve for releasing pent-up anger and frustration.

You can get really worked up during a hearty bout of tickling. Not only do you blow off steam, but the intense, intimate touching can be quite sexually stimulating. Some ticklish areas are also very erotic. When you start playing with them, you're in the twilight zone between tickling and sex. Some tickling can be indistinguishable from foreplay. Furthermore, it's easy to alternate between tantalizing tickles and sex play, with the pitch rising feverishly with each switch.

Tickling can be very effective with women who have problems reaching orgasm. The difficulty often stems from an inability to let go and accept pleasure. The conscious mind won't release control and allow the body to flow with its sensations. Sustained tickling can literally drive someone out of his mind, thus loosening its tenacious grip over the body. Several tickling/sex combination sessions could be the key to overcoming such an orgasm obstacle.

The Tickle Stroke Fingers, of course, are the main tickle tools. But you can't just use them any old way. The pressure should be exquisitely light and gentle, barely brushing the skin. Fingers should be wiggled rapidly over the surface, either at separate speeds and different directions, or together in small, sweeping circles. Without this fast, soft action, a light touch will lose the tickle response and become purely erotic.

The Ticklerogenous Zones People are surprisingly uniform in their reactions to tickling. If you're persistent enough, even the "I'm not ticklish" types usually break down. There appear to be three primary responses to the tickle stroke, which vary according to body area. These "ticklerogenous zones" include: Zone 1 (super-intense); Zone 2 (intense); Zone 3 (ticklerotic). Other body parts than these zones respond to the tickle stroke, but with less passion. Your partner may have his or her own unique "ticklenook" requiring special exploration.

Zone 1 includes the sides, tummy, knees, and the area right above the pubic hair. There are the *top spots*, and with patience even the most unticklish partner will succumb here. If you really want to make your partner writhe, start on one of his sides, travel across the waist (try the navel) and go down one

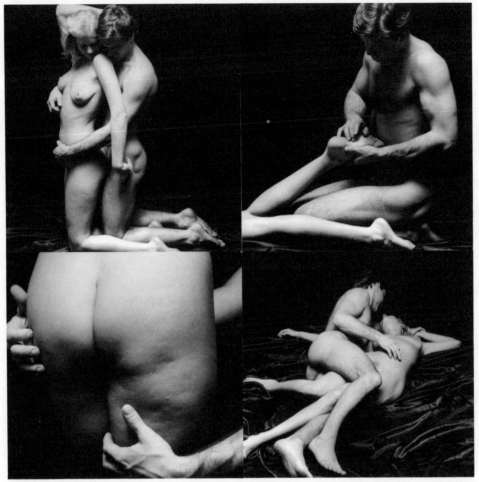

Zone 1 (tummy); Zone 2 (feet); Zone 3 (buttocks); Tickletorture.

of the V-shaped lines that flow from the pubic area. This is bound to get results.

Zone 2 consists of the back, neck, underarms and feet. In most cases, a subtler, longer stroke is required. You can use Zone 2 as "foreplay" for the intenser zones. However, once you waken the body to the tickle response in another zone, Zone 2 livens up as well.

Zone 3 covers many of the usual "erogenous zones." Used here, the stroke can cause an erotic, ticklish, or mixed reaction. Zone 3 is the most intriguing, because it links tickling with sex. There's a lot of variation in this zone, which includes the breasts, buttocks and between the thighs. The female's reaction depends partly on whether or not she's had an orgasm. Before she climaxes, the stroke enhances pure eroticism, and can be an excellent variety of foreplay. It often leads to bigger and better orgasms, after which the tickle response grows stronger. One spot which often has an incredible post-orgasm ticklishness is the clitoris. This sensitive area can become unbelievably ticklish an instant later. You must be really tender here (sometimes it's better to tongue-tickle).

Ticklewrestling Two lovers who have a no-holds-barred contest to see who's the best tickler are "ticklewrestling." It's a pretty harmless outlet for stored-up hostilities, whether they're the mundane ones that build during a workweek, or the special kind that form in a relationship. The object is to make your partner surrender first. With two trained ticklers, it can become rather hot and heavy. A warning to men: be prepared to lose, because women often have quicker fingers and a softer touch. However, you won't learn many nicer games to lose! Ticklewrestling can really charge a couple up, a nice preliminary for some satisfying sex.

Tickletorture This is the "hard core pornography" of tickling, and should only be used on special occasions. Practically speaking, tickletorture can only be done male to female. Few women are strong enough to hold the position effectively on a male partner. It can be most useful with women who are uptight about climaxing. In most tickle play, the female can use countermeasures. She can tickle back, bounce off the bed, grab the guy's hands, or just leave. She can't do any of that with tickletorture. You can literally drive her out of her senses.

If you want to try this harmless little dominance game, don't tell her your intentions. Play it sly. Place yourself at one of her sides, say the right. Have her slip her right arm around you while you put your left under her neck. To avoid cramping, be sure both your left and her right arm fall in between your bodies and the pillows. Keep just enough pressure on her right arm to hold it underneath you. Then tenderly place her free left hand in your left, and hold it gently but firmly. At about the same time, slide your left leg under her legs, and your right leg over. You'll then be able to lock her legs in between, in a sort of soft semi-scissors grip.

This may sound difficult to playfully dupe her into, but it really isn't. You simply make it seem that you're searching for a cozy position to cuddle her. You appear to be holding her softly, then suddenly you tighten up and she's caught. The key is to apply the various grips simultaneously. This keeps her off guard. At one point, you're lovingly holding her left hand, and your legs are pleasantly wrapped around hers. The next instant, she finds herself trapped! You now have a free right hand to do anything you want. She'll beg, plead and bounce furiously once you start tickling. Tantalize her by switching between fast and slow strokes. Thrill her with occasional sexual touches. Unless you're Samson, she'll eventually bust loose. But, hopefully, not before you've tickled her out of her wits. From there, the possibilities are endless.

In some respects, this entire book has been about sex. All the touch techniques have one thing in common: they put you in greater contact with your own bodily processes. This enhanced awareness can intimately affect your attitudes, as well as your approach, towards sex. By experiencing the wonderful pleasures of the various touch techniques, you learn to love and appreciate your body—and your partner's—in new ways. You become less rigid and more relaxed. Your ideas about bodily intimacy become greatly expanded, leading to a more accepting, uninhibited view of sex.

We've also looked at many methods for enhancing sex. Cuddling, tickling, massage—just to name a few—can all be easily incorporated in a sexual session. Practicing just about any of the touch techniques regularly with a partner will also lead to a special closeness and appreciation between you. You'll know each other's bodies very well, and you'll have grown accustomed to sharing physical delights. This will carry over into your erotic activities. You will more willingly cater to each other's needs, since you're already set a pleasing pattern for doing this. Sexual touch will become an extension of the warmth and caring you've already been sharing.

Few can delve deeply into touch techniques without undergoing a major overhaul in sexual attitudes. The prevailing pattern of sex fulfillment is simply too much at odds with an enlightened view of touch. The so-called sexual revolution did little more than make orgasm the Great National Quest. The rest of the body became sadly neglected in the frenzied hunt for total fulfill-ment. Orgasm is an important, valuable experience, but orgasm-obsession forces you to focus on genital stimulation to the expense of everything else. Genital play may be the climax or conclusion of a sexual encounter, but it should not override the importance of all that occurs before. If touch tech-niques teach any one lesson, it's that the entire body is capable of experiencing gratifying pleasure.

Orgasm-obsession also leads to a harried, rushed frame of mind. After all, if it is the be-all and end-all of sex, why pay much attention to preliminaries? Women seem to be increasingly dissatisfied with this attitude. Questioned for the *Hite Report* on female sexuality, they stressed how crucial body contact and closeness were for them in sex. "Sex should be sheer luxuriating in pleasure, being close to the other person, enveloped in warmth and touching all over— not a race for orgasm," was how one woman put it in the *Hite Report*. Others admitted engaging in sex simply to receive a much-needed male embrace.

If you use the touch techniques with sex, you'll never have to worry about depriving your partner of human contact. The sensual techniques of cuddling, tickling and massage can all serve as sexual foreplay. They can be used to extend a sex session, or to arouse you again afterwards. Even if you don't use them in sex play, they help foster a slower, more refined approach to sex. When you're used to receiving full-body pleasures, you will likely pay more attention to preliminaries. Another possibility is to slow down the pace of sex itself. For example, such Eastern forms of erotic activity as Tantric Yoga use intercourse positions which permit little motion. The sex energies are allowed to simmer and deepen, which often sends the couple into mystical states of ecstacy. By delaying orgasm, sexuality can be transformed into a spiritual experience, a divine union of male and female.

In *Nature, Man & Woman*, Zen philosopher Alan Watts speaks of "contemplative sexuality" in which orgasm-obsession is put aside for more relaxed, spontaneous expression. For example, a couple can simply lie motionless with the female astride the male. With no effort, this sustained contact will eventually produce a spontaneous erection in the male. The female's sexual juices will also flow naturally, so the penis may eventually enter the vagina with minimal movements. After penetration, the couple should continue to lie motionless, letting the sex energies well up deep inside. This position is a perfect marriage of sex and cuddling. After awhile, the orgasm urge will appear — on its own. Nothing will have been done to force the issue. Because of its spontaneous nature, it'll probably be more potent than ever.

Contrast this with the hard-driving, grasping sex so many people now prefer. Some couples will literally tear each other apart in a furious effort to attain the elusive "simultaneous orgasm." Gentleness gets tossed aside, and egos get all wrapped up in this quest. While orgasm-obsession is bad, simultaneous orgasm-obsession is even *worse*. Whatever the wonders of coming together, the headaches simply aren't worth it. It's little more than another egotistical "crowning achievement" for lovers to aspire to. This difficult ideal makes people quite insecure when they "fail" to attain it. The simultaneous requirement can totally screw up spontaneity. The woman wonders, "Am I taking too long?" while the man worries, "Will I come too soon?"

A much healthier attitude is *live and let live*. Regularly using touch techniques with someone helps foster this frame of mind. In most of the techniques, you take turns giving and receiving. For example, just thinking of the notion of "simultaneous massage" shows you how silly excessive mutuality can be. It's impossible to fully concentrate on your pleasure and your partner's at the same time. If people practiced a give-and-take approach to sex, many so-called sex hang-ups would vanish overnight. For example, the "vaginal orgasm" issue. Some women simply can't receive adequate clitoral stimulation during intercourse. Instead of feeling inadequate or spending a bundle on therapy, why not just engage in intercourse for his orgasm? He, in turn, can use alternate means before or after intercourse to help her — vibrator, oral or manual stimulation.

The same reasoning applies to the big bugaboo for men, "premature ejaculation." If a man climaxes fast, there are still a number of other things he can do to satisfy his mate. Trying to hold off as long as he can is not only biologically unnatural for a man, but a positive killjoy for future spontaneous expression. It injects a self-consciousness totally at odds with the naturalness of sex. For most men, the *first* orgasm in a session is the *only* one in which he climaxes quickly. Unfortunately, most "premature ejaculators" feel so devastated after the first effort that they don't try again that same session. A much healthier alternative would be to freely accept the first quick climax, then have the male use alternate means to sate his partner. Touch techniques or petting can be used to slowly stimulate him to full potency. From there on, sustained intercourse should be much simpler.

The main requirement for good sex is that each partner be satisfied in a way which appeals to him or her. It should make little difference *when* each partner climaxes, as long as it takes place some time during the session. Nor should it matter whether the method of climaxing is considered "proper" in sex technique books. As long as no one is harmed, what difference does it make? Too much sex play is viewed nowadays from a judgmental perspective, with "performance" and "standards" constantly kept in mind. Those who are different or don't measure up are made to feel inadequate.

Touch techniques help increase acceptance of diversity because they show you the wide variety of thrills the entire body is capable of yielding. It's hard to condemn someone else's sexual proclivities after experiencing several of the unusual pleasures that such unorthodox stimulations as foot reflexology can give. If you share touch techniques with a number of friends, you'll also find that different people are drawn to different methods. Some people prefer Swedish massage; others love Shiatsu. Psychic healing may be one person's cup of tea, while Touch For Health may suit another. You come to realize how individualistic people's bodily desires are, and you can't help but feel the same way about sex preferences. This helps build a non-judgmental "anything goes" attitude toward sex.

For example, let's say you're a man who meets a woman whose only satisfaction comes from oral stimulation of the clitoris. You don't mind doing that, but you really want to have straight intercourse for your own needs. There's no reason for conflict here if each partner is willing to satisfy the other in the way he likes best. Problems only crop up when you start fretting about "simultaneous orgasm," "proper standards," "deviance," "performance," and similar nonsense.

Now that we've examined the effect touch techniques can have on sexual attitudes, it's time to shift to the relationship between sex and touch. Sex can be seen as a sort of "touch overload." By intensely stimulating erotic tissues, we increase the pleasurable touch-reaction until we produce too much heat to handle. We then get a "circuit-breaker" effect, the orgasm, which rids us of the excess energy and brings us back to a low-stimulation state. The orgasm is the body's ultimate response to the ecstacy of touch. Sex extends the touch experience in another significant way. Unlike other bodily intimacies, intercourse involves the interpenetration of bodies. The boundaries separating the partners are broken down. The woman is stroked inside by the male organ. For the male, a part of his body becomes engulfed within the female frame.

Female responses to sex are more complex than the male's. Genital stimulation for a woman take two forms: external (clitoral) and internal (penetration). The male is limited to external stimulation of the erectile tissue of the penis. The male generally feels greater pleasure as his penis is more tightly and completely enclosed inside flesh. The female learns to enjoy two types of touch. Intense penetrating movements can be quite gratifying (though not necessarily orgasmic), while subtler, sensitive touching on or around her clitoris can produce ultimate joy. One highly (and unfairly) maligned act which intensifies both male and female sexual touch is anal intercourse. The male's shaft receives the tightest bodily engulfment possible, while the female experiences the deepest, most intense penetration.

Anal sex has a bad reputation for many reasons. (Indeed, it is illegal in most communities.) Faulty technique has turned off many a woman. To be pain-free, it's essential to relax the area first with tongue and fingers, and to use an abundant amount of lubricant like oil or vaseline. Many people consider it messy, though that can also easily be remedied beforehand. Others associate it with homosexuality, a somewhat specious conclusion when it's being done with a woman. How can any act a man performs with a woman be "homosexual"? Finally, there's the persistent social horror about anything linked with human waste. Even in the age of sexual liberation, the anus remains the last major Sex Taboo. Which is too bad, because the area contains many sensitive pleasure nerves. Done properly and gently, many women learn to relax and enjoy the unique penetration of anal intercourse. The man's penis, in turn, receives a touch treat which is hard to match in any other spot. The position can be combined with clitoral stimulation for complete satisfaction for both partners.

Anal sex aside, the main thing to remember about sexual touch is this: it should be supplemented by as much non-sexual contact and closeness as possible. The body should be treated as one continuous erogenous zone, with no part deserving to be left out. Adding intimate touch techniques to your sexual repertoire will immeasurably enhance your experiences. You'll be satisifed in a fuller, deeper way. The joy of touch can most definitely take the joy of sex to new heights.

BIBLIOGRAPHY / REFERENCES

PART I: The Touching Drive
Eibl-Eibesfeldt, Irenaus. *Love and Hate*. Holt, Rinehart and Winston, New York, 1971.
Hite, Shere. *The Hite Report*. MacMillan, New York, 1976.
"Love that Linus-ism," *Psychology Today*, January 1980, p. 96.
Lowen, Alexander. *The Betrayal of the Body*. Collier, New York, 1969.
Montagu, Ashley and Floyd Matson. *The Human Connection*, McGraw-Hill, New York, 1979.
Montagu, Ashley. *Touching: The Human Significance of the Skin*. Harper & Row, 1971.
Morris, Desmond. *Intimate Behaviour*. Random House, New York, 1971.
Simon, Sidney. *Caring, Feeling, Touching*. Argus, Niles, Il., 1978.

PART II: Touching Tools of the Body
Lee, Linda and James Charlton. *The Hand Book*. Prentice-Hall, Englewood Cliffs, N.J., 1980.
Montague and Matson, *see Part I*.
Morris, Desmond. *The Naked Ape*, Dell, New York, 1970.
Rama, Swami, Ballentine, Rudolph and Alan Hymes. *Science of Breath*. Himalayan Institute, Honesdale, Pa., 1979.
Villee, Claude. *Biology*. Saunders, Philadelphia, 1964.
Weber, E. H. *The Sense of Touch*. Academic Press, New York, 1978.

PART III: Touch Techniques: Stimulation and Relaxation
Gunther, Bernard. *Sense Relaxation*. MacMillan, New York, 1968.
Hofer, Jack. *Total Massage*. Grosset & Dunlap, New York, 1977.
Inkeles, Gordon. *The New Massage*. Putnam, New York, 1980.
Levy, Ronald. *I Can Only Touch You Now*. Prentice-Hall, Englewood Cliffs, N.J., 1973.
Simon, Sidney, *see Part I*.
Zerinsky, Sidney. *The Swedish Massage Work Book*. Swedish Institute, New York, 1975.

PART IV: Touch Techniques: Health and Well-Being
Carter, Ruth. *Hand Reflexology*. Parker, West Nyack, N. Y., 1975.
Carter, Ruth. *Helping Yourself With Foot Reflexology*. Parker, West Nyack, N.Y., 1969.
Davis, A.R. and W.C. Rawls. *The Magnetic Effect*. Exposition, Hicksville, N.Y., 1975.
Davis, A.R. and W.C. Rawls. *Magnetism And Its Effects On The Living System*. Exposition, Hicksville, N.Y., 1974.
Davis, A.R. and W.C. Rawls. *The Rainbow In Your Hands*. Exposition, Hicksville, N.Y., 1976.
De Langre, Jacques. *Do-in 2: The Ancient Art of Rejuvenation Through Self-Massage*. Happiness Press, Magalia, Ca., 1978.
Eden, Jerome. *Animal Magnetism and the Life Force*. Exposition, Hicksville, N.Y., 1974.
Eden, Jerome. *Orgone Energy*. Exposition, Hicksville, N.Y., 1972.
Florian, Tom. "An Introduction to Polarity Therapy" (mimeo).

Gordon, Richard. *Your Healing Hands: The Polarity Experience*. Unity Press, Santa Cruz, Ca., 1978.

Inkeles, Gordon, *see Part III*.

Krieger, Dolores. *The Therapeutic Touch*. Prentice-Hall, Englewood Cliffs, N.J., 1979.

Mann, Felix. *Acupuncture*. Vintage, New York, 1973.

MetaScience Quarterly, Vol. 1, No. 1, Spring 1979.

Miller, Roberta DeLong. *Psychic Massage*. Harper & Row, New York, 1975.

Oliver, William. *New Body Reflexology*, Bi-World, Provo, Utah, 1976.

Ostrander, Sheila and Lynn Schroeder. *Psychic Discoveries Behind the Iron Curtain*. Bantam, New York, 1971.

Oyle, Irving. *The Healing Mind*. Pocket Books, New York, 1976.

Ramacharaka, Yogi. *Fourteen Lessons in Yogi Philosophy*, Vol. 1. Yogi Publication Society, Chicago, 1931.

Rueger, Russ, "Tripping The Heavy Fantastic," *Human Behavior*, March, 1973.

Schultz, William. *Shiatsu: Japanese Finger Pressure Therapy*. Bell, New York, 1976.

Serizawa, Katsusuke. *Massage: The Oriental Method*. Japan Publications, Tokyo, 1980.

Teplitz, Jerry with Shelly Kellman. *How To Relax and Enjoy*. Japan Publications, Tokyo, 1977.

Thie, John. *Touch For Health*. DeVorss & Co., Marina del Rey, Ca., 1979.

PART V: Touch Enhancers

Braun, Saul, ed. *Catalog of Sexual Consciousness*. Grove, New York, 1975.

DeBold, Richard and Russel C. Leaf, eds. *LSD, Man & Society*. Wesleyan University, Press, Middletown, Connecticut, 1967.

Frazier, Gregory and Beverly Frazier. *The Bath Book*. Troubador Press, San Francisco, 1973.

"The Great Playboy Sex-Aids Road Test," *Playboy*, March 1978.

Grinspoon, Lester and James Bakalar. *Cocaine, a Drug and its Social Evolution*. Basic Books, New York, 1976.

Johnson, Tom and Tim Miller. *The Sauna Book*. Harper & Row, New York, 1977.

Maclay, K.T. *Total Beauty Catalog*. Coward, McCann & Geoghegan, New York, 1978.

Lilly, John C. *Deep Self*. Simon & Schuster, New York, 1977.

National Institute of Mental Health, "Students And Drug Abuse," *Today's Education*, March 1969.

Psychedelic Review, Vol. 1, No. 2, Fall 1963.

Psychedelic Review, Number 6, 1965.

Psychedelic Review, Number 10, 1969.

Rueger, Russ. "Postscript To A Bum Trip," *Human Behavior*, November 1973.

Swartz, Mimi, "For The Woman Who Has Everything," *Esquire*, July 1980.

PART VI: Touch Techniques: Intimacy

Braun, Saul, *see Part V*.

Comfort, Alex, ed. *The Joy of Sex*. Simon & Schuster, New York, 1972.

Dement, William C. *Some Must Watch While Some Must Sleep*. Norton, New York, 1978.

Hite, Shere, *see Part I*.

Monroe, Robert A. *Journeys Out Of The Body*. Anchor, New York, 1973.

Rueger, Russ, "The Art of Erotic Tickling," *Gallery*, October, 1976.

Rueger, Russ, "8 Ways to Win and Keep a Super-Beautiful Woman," *Gallery*, August, 1980.

Rueger, Russ, "The Joy of Anomie," *Human Behavior*, July 1974.

Rueger, Russ, "Seeking Freedom from the Male Myth," *Human Behavior*, April, 1973.

Rueger, Russ, "The Sensuous Art of Cuddling," *Penthouse*, August 1973.

Watts, Alan W. *Nature, Man and Woman*. Vintage, New York, 1970.